General Sosabowski's Tourist

General Sosabowski's Tourist
A Polish Paratrooper's Memoir

Boleslaw Ostrowski

WALKA BOOKS
NEW ORLEANS

Walka Books
2713 Royal Street
New Orleans LA 70117

Copyright © Boleslaw Ostrowski

ISBN 10 : 0972290311
ISBN 13 : 978-9722903-1-9

Cover Illustration By Magda Boreysza

Photographs Copyright ©
Author's Collection
Publishers Collection
Polish Institute and Sikorski Museum, London

This Edition Copyright © Walka Books

walka.books@gmail.com

This Book is Dedicated to the memory of my comrades of the Signals Company, 1st Polish Independent Parachute Brigade, and to the people of Holland who experienced untold suffering during the Battle of Arnhem, but never blamed us for the destruction we left in our wake.

To the contrary, they have opened their homes to us, and continue make to welcome us as the had when we had first "dropped in" 70 years ago.

Boleslaw Ostrowski
Mississauga, Canada

Part 1

Dubno
Volhynia

I begin this memoir with the few memories I have of the Ostrowski family history. The Ostrowskis were members of the Polish nobility which had lived in Poland's eastern borderlands for centuries. In these borderlands, known to us Poles as the "*Kresy*," our family had most likely had a significant social role, but as a consequence of their participation during the Polish struggles for Independence through the eighteenth and nineteenth centuries, had lost both wealth and their lands. Entire generations of Poles living in Volhynia and other areas taken by Tsarist Russia during the course of three partitions, had suffered bloody persecutions and deportations to Siberia. Assiduously kept manuscripts and various family mementos had been destroyed during the First World War, which had ravaged the *Kresy*. My grandfather Julian spent much time writing, trying to restore from memory these manuscripts, but this legacy came to an end with his tragic death.

I know my grandfather Julian and my uncle Antoni, my father's brother only from old family stories. The Ostrowskis had lived outside of the village of Klewan on the Horyn River for centuries. The estates of Prince Radziwill surrounded Klewan, and my grandfather was a forester on one of them. It was here that he fathered seven children; my father, Stanislaw his brother Antoni, and their five sisters.

Well before the outbreak of the First World War, my family moved to a farm on the Ikva River, some five kilometers from the town of Dubno. It was here that they settled for good. Before the war broke out, my grandfather Julian was already in his elder years. Since my Uncle Antoni had a successful career, my father took on the task of running the family farm.

During the entire period of the First World War, from 1914-18, the southeastern region saw fierce and prolonged fighting. The region began in Russian hands, then came the Austrians, and finally the Germans. All these occupying armies brought frightening destruction with them. All of the soldiers of all of the armies were very aggressive in officially requisitioning food and livestock, while the soldiers of all the armies simply stole. For the aged, the infirm, the women and children, their larders, barns and homes plundered, sought shelter in the dark forests. That was the

1912
My Uncle Antoni Standing,
My Father Sitting, Together With
Their Three Sisters

My Father, Sister Fela, My Mother, Brother Kazik and Sister Bronia Before the War. Bronia Would Not Survive.

sad reality of these years.

The end of the war in 1918 brought our region widespread lawlessness as numerous bands of deserters and thieves ravaged what was little was left in Volhynia, and robbed and murdered the local people. One night bandits burst into our house. They shot my grandfather and uncle out of hand, and then started looting, and smashing what they could, or would not steal. Fortunately, my mother and grandmother managed to escape through a back window, and hid in the forest.

The years 1919-20 were little better, and brought many dramatic and heart-breaking incidents. After the Russian Revolution, and the overthrow of the Tsar, hordes of Bolsheviks began to pour into the *Kresy*, marched deep into Poland, and managed to get almost as far as Warsaw. Finally defeated at the gates of the capital, the now shattered and disorganized Bolsheviks fled in panic, pursued by the Polish Army. The retreating remains of the Red Army stopped only to further plunder what they could. It was only after the new Polish state brought peace to its eastern regions that it became possible for people to even begin thinking about rebuilding their lives.

In the meantime all of my father's sisters managed to marry, and began raising families in the Dubno region. Now alone on the farm, my father could begin to take of, and raise his own family. My oldest sister Felicia was born in 1907. My brother Kazimierz followed two years later in 1909. Another girl, Bronislawa, was born in 1911, but she died at the age of seven due to the lack of medical care in the war-torn countryside.

The outbreak of war in 1914 brought a temporary halt to any new additions to the family. But, after my father returned from Austrian captivity sought to change that. I was born in 1919, and my brother Dionizy in 1921. During these turbulent years an infant presented an additional problem. My mother would leave me for hours hidden in corner of our coach house's loft, leaving my fate to Divine providence, and praying that I would not to cry too loudly.

When the threat of a raid by marauders was imminent, the entire family sought refuge in a nearby grove, and left our home and belongings to God's mercy. The only exception were our horses, which were especially sought by the rampaging thugs. My older brother Kazimierz, despite his tender age of 10 years, was a very brave and resourceful boy. During unexpected visits from robbers or deserters, Kazimierz took care of the horses, bringing them to the thicket, where he remained hidden with them, and kept them quiet.

Our farm was located at a crossroads. During the daytime it was very easy to see who was approaching from a distance. But it was a different story at night, and my father did not sleep in the house during the summer months. Instead, he had a cot in the coach house. After saying his nightly prayers, and reciting a long litany to St. Stanislaw, my father (who was named Stanislaw, and kept the saint as his patron) then turned in with a loaded double barreled shotgun next to him.

Our farm was also guarded by dogs, which always let us know if anybody was approaching. Our isolated farm was a Polish Island in a Ukrainian sea. Our nearest Polish neighbor was almost two kilometers away on the main road to Rowne, and we always turned to them for help when needed. But to get to them, we had to run across the fields and then through a dense and thorny thicket.

Not Bolshevik Hordes -
But Soldiers From The Local Garrison Commemorating The 1920 war

Most of all we relied on our friends the Cherepoviches, who despite the fact that they were Ukrainian, understood our position as Poles. Our relations with them were always very warm. They were true friends whom we could always depend on. Mr. Nikolaj Cherepovich worked as a game keeper in the nearby forest, and as a result knew each and every person and every tiny corner in the region. My maternal grandmother was especially attached to them. After my grandfather and uncle were killed, Mr. Cherepovich tirelessly searched for their murderers. Despite the fact that he managed to track them down, there still was no functioning legal system, and they escaped justice.

DUBNO
The City Of My Youth During The 1920s

The 1920s were extremely difficult for the newly reborn Polish Republic. The young nation found Itself in a state of constantly changing governments which could not contribute to any sort of economic or political stability. The map of Europe had changed, with newly independent nations emerging from the breakdown of the three defeated empires. Serious problems also emerged with the smaller states suffering conflicts with their larger, more powerful neighbors. In Poland's case, our country was between Germany and Soviet Russia. There was also a very large Ukrainian minority in the southeastern part of Volhynia. A radical nationalist minority, who sought an autonomous Ukranian part of the so-called

Ukrainian Unification began an underground program of violence against the Poles, and even murdered moderate Ukrainian leaders who opposed their merciless and bloody methods. Murder brought retaliation, which helped to polarize neighbors. When the Second Poland Republic was approaching its tenth year, the entire world was plunged into an economic crisis, which created a situation for the nation that was even more tragic than the extreme difficulties suffered by the developed countries.

But by and large, life outside of Warsaw, and especially in the faraway *Kresy*, carried on as it had before the war. The Tsarist regime, until the outbreak of the First World War, cared little for old Polish borderlands, where the cities towns were usually Polish and Jewish, while the villages were mostly Ukrainian, along with scattered colonies of Czechs or Germans. The rural regions carried on as it had before the war, with the usual economic shortcomings that left them impoverished. This was all compounded by the years of war, destruction and lawlessness, and as a result, every stratum of society and every nationality had to shoulder the burden.

My father worked hard on the farm from dawn till dark. Slowly everything was beginning to take shape. He rebuilt the neglected apiaries, of which he was very proud, and I, then a six-year-old sprat, was in charge of watching the bees while he looked after many other things. Usually, after leaving the old hive, a new, young swarm would spend two or three hours on a nearby tree, as if to gather strength before taking flight. These two to three

Sister Felicja And Brother Kazimierz, 1922

hours were critical, since during that time the swarm of bees were suspended into in a large mass that resembled a big ball. It was necessary to recognize the queen bee, and put her in a special box with some fragrant mint leaves. Then the swarm would follow the queen bee without any problem to a new hive.

Apart from the few odd duties delegated to me by my father, I always found plenty of time to play. My brother Dionizy, who was two years younger than me, tried to be like me, but could not keep himself out of trouble. Despite his mischievous antics, we passed carefree boyhood.

I remember that winters in Volhynia were very cold, and that we had unbelievable amounts of snow. Today, winters are very mild, but I can never forget the winter of 1928 – 1929, when life almost came to a complete halt. One morning we awoke from a blizzard which had howled through the night. It turned out that so much snow had accumulated that it was impossible to leave the house. We were on a farm and of course we had to take care of our horses, cattle, and chickens. To do so, we had to dig tunnels from our house to the outbuildings. My father would bring the healthiest and most populated bee hives into a heated utility room which was attached to the house. The hives that remained outdoors would not survive the winter.

All of the snow and ice accumulated during the winter would thaw during the spring, and flow into the low-lying areas and create bogs and ponds. One of these was like a small lake for many months of the year. Surrounded by reeds, it had a small "island" in the middle. It was a place where my father liked to go duck hunting.

When our parents were out of sight, my brother and I "borrowed" my mother's washtub, because we did not have a boat. Fishing in the pond was impossible, because, of course, there were never any fish there, so we hunted frogs in the reeds. Bring dozens of them home, we were treated by a unique concert by the frogs that evening. My parents were not amused, but to me it was such a wonderful performance that I had have never heard anything else like it in my life.

Among the other youthful adventures that we youngsters took part in were observing various birds, watching them nest, and gathering sparrow eggs. The sparrows, of course, were not very happy with this, but at times my father cursed them for the damage they did to our barn's thatched roof.

In the spring, the storks would return to Poland after spending the winters in Africa. They had their nest on our barn, and as soon as they arrived they immediately began repairing the damage their brooding home had suffered during the long winter.

Storks were large, but very shy birds, and were welcomed as

they were considered to be good luck. But Dionizy could not let his curiosity about them rest. He just had to see what the inside of their nest looked like. He climbed up a tall pear tree next to the barn and
way to the roof peak. As we all knew, storks, though shy, are very protective of their nests, and would vigorously fight any intrusion, especially if their were any young in the nest. The stork went after him with his long sharp beak, and my brother fell off the roof and hit the ground in a cloud of dust. I thought that he had broken his neck, but this adventure taught him nothing, and he continued do things his way.

    To be able to imagine how mischievous Dionizy was, I will relate a few of his better "ideas." These usually came to him when our parents were not around, and became some of the greatest worries of our grandmother, who was usually looked upon as the supervisor for the entire farm. In one instance he tried to catch a rooster. This particular rooster was the most active in chasing the hens in the farmyard, for the precise reason that roosters pursue hens. His amorous efforts left the poor hens tired, tormented, and missing a lot of feathers. Dionizy tied an old tin can to the rooster's tail, and the bird ran around the yard as it was on fire, until it keeled over, dead. When grandmother saw the rooster and asked what had happened to it, Dionizy quite convincingly told a story about how he had seen a large hawk swoop out of the sky and sink its talons into the rooster.

My Brother Dionizy
1938

    Some of Dionizy's excesses could have ended tragically. These usually happened when our parents were not watching. Grandmother usually took a nap after her afternoon meal. Dionizy was in the front room, playing with something while I was looking at pictures in a book about life during the times of the Old Testament. Suddenly there was a loud explosion, as if lightning had struck the house. I just about shot out of my chair while my grandmother, suddenly awakened by the explosion, actually fell off of the couch onto the floor, clutching her heart.

    What was the cause for the explosion? It was my father's shotgun, which had its place on the wall, and as it always had been during past days, it remained loaded and ready in case of the arrival of unwanted "visitors." Dionizy knew very well that it was not allowed, under any circumstances, to take the gun off of the wall. But lying down, he was touching the shotgun with his feet, and eventually, he even managed to cock the hammer. What followed was a lesson that Dionizy will carry with him for the rest of his life. The jagged hole in the ceiling, the cloud of gunpowder, and Grandmother's moaning laments told my father the entire story. Father took off his belt, and applied it with an energy that Dionizy had never experienced before. Would this cure him of his mischief? We all had our doubts...

Student Days

In fact, Dionizy had a good heart and a unique gift, which he still possesses today. Despite being in his waning years, and not in the best of health, he continues to entertain everyone with his funny stories and jokes, frequently the creations of his own fertile imagination.

The war years separated us for decades, he in Poland, and I in Canada. Whenever I visit my motherland, I always visit him. He lives with his wife Adela in Cigacice, a tiny village on the Oder River, north of Zielona Gora. The couple raised chickens, and worked the small piece of land around their home for most of their needs, but they no longer have the strength. I ask the reader to excuse the divergence, but these stories about my brother Dionizy help to illustrate what life was like for us growing up during the 1920s.

When I turned seven, it was time for me to start school. I went reluctantly, especially as I was enrolled in the nearby Ukrainian school in Raczyn, where everything was conducted in that language. Such schools existed during the early years of the Polish Republic.

Dubno was my home town, and the county seat located on the Ikva River in Volhynia. It was well-known for the castle built by the noble Ostrogski family during the late 15th Century. The castle's fortifications repulsed the Tatar hordes in 1577, and withstood a long siege by Cossacks during Bogdan Khmelnitsky's revolt in 1648.

When I was growing up, the town of Dubno had a population of some 20,000, of which half were Jews. Despite the fact that most of the trades, businesses and the small factories were owned by them, the majority of the Jewish residents lived in extreme poverty. Much of this had to do with Tsarist policies which established the "Pale of Settlement." Jews who did not fit into certain categories, such as wealthy merchants, those who had either special or higher education, skilled craftsmen, engineers, or, who in general, were not considered an asset to the Russian Empire were re-settled in the territories seized from Poland during the 18th Century partitions, and this example of "ethnic cleansing" went on almost to the dawn of the 20th Century.

Dubno was the county seat, with municipal offices located in the castle. There a cathedral, three other large churches, both Catholic and Orthodox, several synagogues, a gimnasium, and a business college on Panienski Street. The town was also home to two military units, the 43rd Infantry Regiment, and the 2nd Horse Artillery Battalion which had their garrisons there.

I lived in town despite the fact that I was born on an small farm some five kilometers that my family had worked for years. I spent my summer vacations from school on the farm until I began attending commercial school, and then gaining practical experience working for Mr. Szczepaniak in his large store.

The Unforgettable September Of 1939

And so Dubno, as a frontier town, far from the center of the nation and its capital, lived a quiet and monotone existence, even on the day that the war broke out. But during the days that followed, when the massive German armed forces with their "Blitzkrieg" smashed through the Polish defenses, refugees began streaming into Dubno. The stream soon became a river of exhausted and frightened humanity, mostly from the western parts of Poland They presented a sad sight, carrying their bundled possessions, traveling on foot, or riding or pushing anything that had wheels, often accompanied by livestock. They were followed by bombers with black crosses on their wings, whose whining engines created fear which felt like a drill auguring through one's bowels, and that fear would turn to shear terror when the bombs started screaming and exploding.

We began to realize the horror of our situation, with the German bombers returning destroying the railroad station. Our town was packed with hungry and frightened people, who had no idea where to go, or what to do. It seemed that nothing would stop the Germans, and we dreaded the loss of our motherland and our freedom, and knew we were facing life under occupation.

But German soldiers never entered Dubno. Rather, our freedom was blotted out by a massive invasion of Soviet armies which started pouring into Poland from the east on September 17. During those days nobody had the slightest expectation that our eastern neighbor would become part of this disgraceful enterprise, and would stab us in the back. Together with Nazi Germany, the Soviet Union and would be initiating the Fourth Partition of Poland.

Dubno Under Soviet Occupation
1939-1941

Life under Soviet occupation was unbelievably difficult, especially for the Poles. The food supplies harvested or stockpiled before September were gradually consumed, and the occupiers were not concerned in the least, as they were busy robbing everything else themselves. As all of the shops had been closed down, and there were no longer any market activities, we had to go out to the neighboring villages on our own to try and find food however we could.

I had been working as an apprentice at the Szczepaniak's shop since 1937, I had every intention of eventually opening my own business. Mr. Szczepaniak had been called-up by the army during August, and his wife Jadwiga and I together tried to exist. We quickly realized that everything eventually would be sold, and the opportunity to get the most basic items, such as kasha, flour and sugar would become impossible, as there was no other supply to be had.

The big question was what were we to do to survive? We had all ready taken the best, most expensive items, and put them into a storage area. In the end, we decided that we should rescue what we could. With the permission of our pastor, Father Kuzminski, I dug a large pit behind his rectory. Over the next few nights, I removed the best vodkas and liqueurs, and buried them for safe keeping in the pit. At that time, we still had the hope that things would eventually return to some sort of normality.

During this entire period, from the moment the Soviets had entered Dubno, to the beginning of mass deportations during the winter of early 1940, we knew no peace. The nights were the worst, and I would often hear gunfire in the streets, or the sounds of people being arrested or interrogated. As a result of denunciations by Soviet sympathizers, or by those who hated their Polish neighbors, many of us chose to remain in our homes, behind tightly drawn window hoping to keep a lower profile and lessen our exposure to the occupiers. I remember that we spent most evenings with the Szczepaniak's listening to the BBC on one of the few radios that had not been confiscated by the occupiers.

While I continued to look for some kind of work, my family had to deal with difficult financial situation. Several years before the war, my family had decided to move closer to their daughter Felicia, in the small village of Bortnica. This village was populated by settlers from Poland, one of which one was my sister's husband. The majority were veterans of the 1919-1920 war between newly reborn Poland, and a two-year old Soviet Union. Those veterans who had served honorably were given the opportunity to settle on a 12 hectare parcel in the Polish borderlands. When one of the settler's parcels became available, my father decided to buy it. He sold the family farm during the spring of 1939, and deposited the money in the bank in order to secure the land. By the end of September of that year, both the bank and every penny that my father had made during his life were both gone.

Bortnica

We were most fortunate that my father was known for his honesty, and was well-liked within the parish. After some discussion, our pastor, Father Kuzminski managed to find accommodations for my family among the parishioners, so at least they had a roof over our heads. Little did I know that the loss of our family fortune would have very little meaning compared to the events that we would all be facing soon.

My situation was also changing. The Szczepaniaks, who were originally from Poznan, in western Poland, were able to return. During this early period of Poland's Fourth Partition, the Nazis and the Soviets negotiated repatriation of citizens whose origins were now in each other's territory. The Szczepaniaks managed to return to their old hometown, and German rule after that city was annexed to the Reich. And in this manner I not only lost some very, very good friends, but also the opportunity to learn the ins and outs of business. We were separated for many years, until the first time I returned to Poland after the war. It was 1972, and we once saw each other in Poznan as very old, and very dear friends.

During the late autumn, our occupiers opened a teachers college and introduced basic courses for teachers. That profession was considered especially suspect by the Soviets, and the majority of the educators who had taught before the war had been either been arrested or deported. I signed up for these courses, and I began a new chapter in my life.

The first beginning course (in either Ukrainian and Russian languages, naturally) was held in the former St. Konarski gimnasium, which had been closed by the Soviets during the third month of occupation. By the end of the winter, the entire teachers school was transferred to Ostrog, some 45 kilometers to the east, and reopened in the former pedagogical lyceum there. I was very fortunate that many of my classmates had also signed up for the course, so there were familiar faces around me. I realized that there was now little chance for any further education, and a way to eke out a living in the new world that I was now trying to survive in, so I became a "pseudo teacher," which is the only way one could describe this fictitious absurdity so typical of Soviet life.

DUBNO
JULY 19, 1940

After I had returned from Ostrog, I readied myself to work as a teacher in the primarily Ukrainian villages around Dubno. Under the Soviets Dubno lost its status as the county seat, and became just another town.

Before the beginning of this school year I was informed that I had a job teaching and was to report to the education inspector in Werba for a teachers conference where we would receive their assignments and other information. I was assigned a position in a four-year elementary school in the village of Mykiticzach. This was a large Ukrainian village halfway between Dubno and Werba, and just a little way off of the main road.

I had never been in Mykiticzach before. I was assigned to live with the Machnicki family, and when I arrived at their house there was a pleasant surprise waiting for me. It turned out that for some years before the outbreak of war, the Machnickis had been teachers in the village school, and they were both very qualified and experienced. I was overjoyed that I would be able to benefit from their experience. I was given a small room, where I settled into with ease. I had my meals with the couple, and at that moment my life was very satisfactory.

I was assigned the third grade as I was fairly fluent in Ukrainian, and even in the earliest part of my new career as a teacher I managed to do well. The curriculum was basic: arithmetic, Ukrainian language, elementary geography and science, gymnastics, games and handiwork. It followed that the books used were in Ukrainian, and published by the Soviets.

Looking at my situation, I realized that I lacked practical experience, and experienced difficulty with course preparation, but was saved by the Machnickis. Mrs. Machnicka, who referred to herself as Madame de Bibersteyn-Machnicka, was older then her husband by a good several years. She was a spirited, cultured, and a well educated and a very astute woman. This very kind lady came to my help quickly and very happily. I can only wonder what happened to these wonderful people after June 22, 1941, and when the region found itself under German occupation, and the local Ukrainian nationalists began burning the Polish settlements and killing the remaining inhabitants. If the Machnickis did not hide in Dubno, they might have become victims. I later learned that my good friend Lech Witkowski, who was a teacher in the nearby Ukrainian village of Swidowiec was among those murdered.

But my tenure in Mykiticzach was very short as immediately after Christmas, 1940, I was transferred to the neighboring Czech village of Turkowicze. I was re-assigned because the village's regular teacher had been ill for a long time, and there was little hope that he would return to his post. I was given a small room in the teacher's apartment, which was located in the schoolhouse. I had my meals there, and I must say, that the wife of the ailing teacher was also Czech, and a wonderful cook. Right away I knew that I would not starve while under their roof. After school hours, I did housework in order to help the wife of the bedridden teacher.

It did not take me long to get to know the local inhabitants, and all welcomed me warmly. This was a great contrast with to the cold and suspicious manner the Ukrainian villagers had treated me in Mykiticzach. It required no effort to coexist with the inhabitants of this modest Czech settlement, and it was so easy to win their trust.

A FAREWELL TO POLAND

Unfortunately, this idyllic life was not to last. At the beginning of March I was to report to the medical commission in Werba for an preliminary medical examination for service in the Red Army. The exam was short and sweet. I was well aware of the stipulations of the Geneva Convention which prohibited conscripting citizens of an occupied nation into the occupier's armed forces, but to who was I to complain to? I knew what would happen if I tried to escape to Hungary. The thought of putting my family in jeopardy was impossible, and I became resigned to my fate. There was no way out of having to serve in the army of the enemy of we Poles. So it happened that I was to share the fate of my father who had served in the ranks of the Russian army during the First World War. And so the same fate that had befallen the previous generations of our forefathers from the time of the partitions of our nation over 160 years before was about to befall me.

I did not have much time to prepare, because I had to inform everybody, and especially my family in Dubno. The fact that I would be leaving came as bad news to the residents of Turkowicz because it meant that the school would have to close because of the shortage of teachers.

I had said my goodbyes with my father earlier. He had come to Turkowicz by himself, with a premonition that we would never see each other again. When we parted my father knew well the awful situation I was in. I no longer had the heart to go to Dubno and say my farewells to my mother and brother. The morning I left it was raining. But this did not prevent most of village residents from gathering in front of the school. All the students were there, along with an orchestra. I was very deeply moved at their heartfelt farewell. One of the farmers took me to Werba in his wagon. On the way, we stopped to say my goodbyes at the homes of those who could not come to the village. It was with great sadness and sorrow that I left this quiet and very kind Czech colony.

And so ended my short teaching career as I began my career as a "Krasnoarmiets" (Red Army Man), which indeed was an ironic fate. This was something I would have never predicted in a million years. A passenger train at the Werba railroad station was waiting. I was not alone, there were others in my situation there too, dozens of men my age from the various villages in the area. It was there that I met two brothers, Stanislaw and Witold Lewicki from the Komaszowka colony, which was not far from Dubno. Together we would sharing an existence as vagabonds eking out a miserable existence across the boundless areas of Russia.

I can not remember where we had to get off the train and then board broad gauge railroad cars. Was that in Zdolbunow, or on the other side of the old border in Shepetovka? For the meantime we continued through our beloved Volhynia eastwards. My curiosity and also plagued by my concerns; where are they taking us and what was waiting for us when we got there?

And so fields, forests and villages passed by the open door of the freight wagon we were riding in. The Volhynian landscape is always beautiful, and springtime was beginning to awaken. Snow still covered the fields, but there were places where the bright green spring grass was breaking through the snow, mostly on the sunny side of the railroad embankment.

The whole of this wonderful country, which for centuries had been overrun by invasions disappeared behind us as the clacking wheels of our train unceasingly rolled us eastwards. Yes, this was our wonderful Volhynia - the Volhynia of many generations. We were rooted in the black, fertile earth equally along with, and among Ukrainians, Czechs, Germans, and Jews. Is there anyone who had lived in Volhynia who does not remember the sad verses of the Ukrainian song:

"Volhynia, Volhynia, I will never forget you,
And until my death, I will always remember you..."

Part 2

The Polish-Soviet Border

We crossed the Polish Soviet border sometime during the night. The old pre-1939 Polish/Soviet border still existed, along with Soviet border guards, and continued to exist until the outbreak of the war between Nazi Germany and the Soviet Union in 1941. I presumed that this was to prevent Soviet citizens from getting a glimpse of the different world that existed outside of their borders.

It was very difficult for me to determine exactly when we had crossed it, or where we were, but we now saw significant changes in the landscape. I must confess, I was very curious about how the "Soviet paradise" looked. The subject of our eastern neighbor was discussed very often before the war, and those conversations were mostly negative. As a result I observed everything with great curiosity.

Now the train flashed past decrepit villages and towns, all without any sign of a garden near the houses. We watched as miserable and emaciated cattle grazed alongside the railroad embankments and in small copses. Even through our eyes, we who were young, and always ready to laugh and smile, the contrast was painful: all was a gray dullness and poverty. We were astonished as we rode through the magnificent "black earth" region of Ukraine, which had a reputation for being some of the most fertile and productive farmland on the face of the earth. After the revolution, the independent farmers in the Russian Empire were referred to as a "*Kulaks*," or "wealthy class enemies of the peasants." It took twenty years of Stalinist terror and Soviet management to "dekulakize" the productive and prosperous farms, and turn them into "collective farms." The result was the starvation of millions of people during the well-known famine in Ukraine. That nation was bled dry, and all that remained was misery.

As we crossed the Dnieper River on a steel railroad bridge we were able to admire the beautiful sight of the green awakening of springtime in historic Kiev. But the residents of that beautiful city were gray and shabbily dressed. The dullness and poverty was reflected in their gloomy and blank faces, which was a faded and terribly annoying contrast to the verdant spring foliage.

From the earliest days of Soviet occupation they were beating messages into our heads. Most annoying were the loudspeakers blaring out Soviet songs on the streets of Dubno, especially one that was played over and over:

Vast and Wide Is Our Land...
We Have Many Forests, Fields and Rivers...
I Know No Country So Glorious and Grand...
Where A Man Can Be So Free...

VAST AND WIDE IS OUR LAND...
A Caricature by Mieczyslaw Kuczynski, a fellow "Siberian"

For the duped who believed communist propaganda, I would have advised a visit to this Soviet "paradise," and certainly they would have had their eyes opened.

I was most fortunate in finding the Lewicki brothers, and we became fast friends. At least I had people with whom I could share the "discoveries" in this new universe. As we left Kiev we realized that the train was now traveling in a northeasterly direction. We started to pass beautiful forests, peaceful and quiet stretches that went on for many kilometers where we rarely saw any sign of human habitation.

As we were young, we were not too worried about what was waiting for us. However, most of us were agonized with concern and fear for our families and families at home, and what would to them. From what we had already lived through under Soviet occupation, we were all well aware of how hard one had to struggle constantly, simply to exist.

After two days of riding the rails east, the train came to a halt in a station. The transport's duty officer informed us that we had arrived in Kursk, and that we would be processed here, but first we would be fed. After the long journey from Werba, where we only had what we had brought with us to eat, we were starving. But, we were curious about "Red Army" grub, so we could at least get an idea of what our stomachs could look forward to for the next few years.

From the railroad station we were marched to a military mess hall. It was an enormous room packed with tables and benches, with a corporal at the entrance pointing us to an empty table set with enamel bowls, eight to a side. We sat down and then two baskets of bread were brought with the perfunctory words, "three slices per man." We were so hungry that no one needed a second invitation. I had barely swallowed my first bite of the bread when an awful odor filled the mess hall. Orderlies went down the rows of tables, tossing a salted herring into each man's bowl. The fish were encrusted with salt, and were whole - their heads were still on, and they had not been cleaned. We did not know how to react. Witold held his herring by the tail, pinched between his fingertips, looking ready to vomit.

Looking around, the Red Army soldiers seemed to have no difficulty. They held the fish with their dirty fingers by the head and tail, and gnawed on the fish, stopping only to spit out the bones. An orderly, seeing that we had not touched the reeking fish grabbed them and threw them back into his pot. He apparently felt sorry for us because he returned, and gave us each a piece of salted fatback. We spent the night in barracks.

The following morning we boarded the same train and after several hours we stopped in a small town. The sign on the railroad station announced that we had arrived in Livny, some 400 kilometers south of Moscow.

We were separated into pairs, and were then billeted in private rooms, which reminded me of cells in a monastery, which in fact it had been, and closed as part of the regime's repressions of religion. I believed that our stay here was temporary, and sure enough, we were taken to barracks at the edge of town after a few nights.

Service in the Red Army

There we were issued with the "proud" uniform of a Red Army man consisting of the typical Russian military flannel foot wraps, puttees, breeches, and the pullover shirt known as the "*gymnasterka*." The *gymnasterka* was basically a long, loose fitting "shirt-tunic" patterned on the traditional Russian peasant smock. It dated back to the time of the Tsar, and had a pair of chest pockets, and closed by three buttons below the pointed collar, and had a skirt worn over the breeches, and worn with a wide military belt around the waist. The ensemble was completed by a forage cap, known as a "*pilotka*," which was crowned at the front with a five-pointed red star.

We were well rested after our nights in the monastery as we began learning the predictable military drill. First of all we were broken up into small groups and taught "the march step," which was the way we would march as we were in Moscow's Red Square with Stalin himself gazing down on us from Lenin's Tomb. With our backs straight as arrows, our chests thrust out, behinds tucked in we were slamming our boots on the concrete with every goose step. We also learned the manual of arms, and how to salute as proper Red Army soldiers, and other military customs during the following days.

On May 1, 1941, we awoke to find the barracks square draped with red flags. We were all lined up in ranks by unit in front of a podium, decorated with red banners, and flanked by all of the regimental colors of our division. After the division's band finished played "The Internationale," our regiment's commander called all of the new recruits forward and explained that we were to swear our military oath of loyalty to the People and Government of the Union of Soviet Socialist Republics. Our company commanders then had each of us read the following out loud:

I, a citizen of the Union of Soviet Socialist Republics, joining the ranks of the Red Army of Workers and Peasants, do hereby take the oath of allegiance and do solemnly vow to be an honest, brave, disciplined and vigilant fighter, to guard strictly all military and state secrets, to obey implicitly all army regulations and orders of my commanders, commissars and superiors.

I vow to study the duties of a soldier conscientiously, to safeguard Army and National property in every way possible, and to be true to my People, my Soviet Motherland, and the Workers' and Peasants' Government to my last breath.

I am always prepared at the order of the Workers' and Peasants' Government to come to the defense of my Motherland - the Union of Soviet Socialist Republics - and, as a fighter of the Red Army of Workers and Peasants, I vow to defend her courageously, skillfully, creditably and honorably, without sparing my blood and my very life to achieve complete victory over the enemy
.   And if through evil intent I break this solemn oath, then let the stern punishment of the Soviet law, and the universal hatred and contempt of the working people, fall upon me.

Then all who took the oath signed a scroll, which was turned over to the regimental staff along with our identity books. Our soldiers' books would be returned with the official stamp noting the date we took our oath. The regimental commander then congratulated us on our acceptance as full-fledged soldiers of the Red Army.

After the regiment passed in review, we were given the day off to celebrate the "International Day of Workers." Our midday meal was of better quality and quantity than the usual. But my normal ravenous appetite had abandoned me after I swore "undying allegiance to the Soviet Union and its leaders." When I saw myself in the washroom mirror, I was shocked at my appearance. Had anyone who had known me, and saw me wearing the cursed Soviet uniform would immediately react;"How could you, a young friend, a Pole whose grandfather's great-grandfather was Polish, and with such a renowned family name, how could you even imagine being a "Red Army Soldier," who proudly serves his Soviet nation and ready to give your life for the Great Father Stalin?

Had somebody in 1939 told me that one day I would be wearing the Bolshevik star, I would have thought that he had escaped from an insane asylum. I remembered hearing stories from my father about the "*Branka*," the annual visit by the Tsar's recruiters, which had thousands of Poles conscripted to 25 years of forced military servitude for the Tsar and his empire. I felt depressed and hopeless, but what could I do?

In a short time we were broken up into small groups. This was where we would be assigned to our military specialty. This event was a little bit different from our previous experiences in the Red Army - we were actually allowed to have a choice! I decided that I wanted to be trained in signals, and after completion of training, I would be referred to as a "Red Army *Radista*." I made that the decision because I knew that a signaler's status and daily life would be better than that of a common foot soldier. But I could not know that after June 22, 1941 that there would be a radical change to my plan of going through my service in the Red Army with the praiseworthy title of "*radista*."

As I later discovered that during early May the Red Army traditionally moved the majority of their garrisons from the barracks that provided their winter quarters and went to their "*Letnie Lagiera*" (summer camp), and on May 7, we received our marching orders. We were not privileged to know where we were going and how far we had to travel. All we knew was that we were to march the entire way, and fully equipped and loaded for combat, with the exception of ammunition. Only the company's heavy equipment would carried on horse-drawn wagons behind our column.

All of this might have actually interested me, had the weather been good, but the matter was decided opposite. Despite the fact that it was the beginning of May, and the roads between Livny and Orel taken by our 400th Rifle Regiment consisted of mud and slush because of the spring thaw. As if the cold ankle-deep quagmire which sucked at our boots was not punishment enough, the majority of the march an icy rain poured down on us. In my young life I have never experienced such misery.

The regiment spent its nights on collective farms. If a soldier was lucky enough to able to spend the night in a cottage where there was a fire, then there was an opportunity to at least dry out his overcoat. But such a stroke of good fortune only happened to me once, and like the majority of the soldiers, I spent the other nights in leaky, drafty barns which no self-respecting pre-revolutionary "*muzhik*" (peasant) would have considered suitable for livestock.

The march was exhausting, and it was all one could do to keep putting one foot in front of the other as we traveled down the muddy roads through sheets of rain. When the sergeants started counting off troops for the night's lodging, I was one of the lucky ones assigned to a dwelling, rather than a barn. There was family living there, a father and mother, their two daughters, and a frail, aged man with a long beard who occupied a bed in a dark corner. This family shared a two room cabin. Now, somehow, they had to accommodate three of us - Red Army soldiers.

After deliberating with his wife, the man brought in straw and spread it out on some boards which were just wide enough to sleep on. They helped us with our packs and hung our dripping greatcoats on either side of the stove. The fire was fueled by a combination of straw, dried animal manure, and whatever scraps of wood they had managed to scavenge. The only illumination was from an oil lamp which left most of the hut in darkness.

At any rate, we were relieved at the thought of being able to rest for a night, as I had been certain that I would have collapsed the following day. As my eyes adjusted themselves to the dark as I sat by the weak, smoldering fire, and somehow I began coming back to life. The Russian family treated me with such kindness that warmed me much more than the smokey fire did. I looked around the inside of their shabby home, and that was all that I needed to know about the poverty and misery that these collective farm people lived with.

I could tell that the head of the family was a good-hearted, and kindly man. Despite the fact that I was dead tired, out of curiosity I began chatting with him about his daily life. From what I could gather, they were given an appropriate amount of grain, potatoes, and other produce from the collective. Theoretically, they could sell or barter this to satisfy other needs. However, it was obvious that there was never a surplus of their share after the state took the majority of the harvest.

The man said that they were allowed to keep a small garden, and one cow, but it was obvious that the family could neither afford to buy or feed one. The man told me in a hushed tone that he did think that he could not count on any improvement in his life under the present system because they were unable to even maintain a minimal existence as things were.

The Soviet authorities continued to promise that after the next five-year-plan that everything will be better. These people knew little, except for the lies they were fed by the communist officials. Over and over, they were told that their lives were so much better now than they were during the time of the Tsars. One of the reasons this was widely believed was a result of continued illiteracy among the majority of the peasants.

I slept like a rock that night. In the morning our "host" offered us some boiled water. My overcoat managed to shed much of the water it had soaked up during the march. While it was still damp, it felt luxurious compared to what it was like when I took it off. The man gave me a smile when he shook my hand, and wished me all the luck in the world. I slung my rifle, and joined the rest of my company as it assembled for the day's march.

All of the marching had us moving along, and using the very last remnants of our physical strength. Those who were either ill or exhausted would collapse in the mud and lay there until they were gathered by the carts that followed the marching column. When we finally dragged our tired selves to our final destination, there, ironically, was the commander of the camp with a military band there to greet us at the gate. How could we possibly have entered the camp properly marching to the martial music when we could barely move our feet?

We had marched some 210 kilometers. That was an average of marching some 35 kilometers a day despite the horrible weather. Our summer home was north of Livny, and we were now on the Oka River a few kilometers outside of the city of Orel

The camp, an entire city consisting of tents, was home to a whole rifle division, along with other smaller units. One side of the camp was bordered by the Oka River, which was much larger than our Ikva River in Volhynia. On the other side of the camp was a large empty area which stretched as far as the eye could see. This was our training area with firing ranges for both infantry and artillery.

I thought about our six day journey, and how, now that it was over, we were really going to get it in the neck here. It was a large camp with a captive population, but I had to acknowledge that I was impressed with the order and cleanliness of the place, which was exemplary. We lived with eight to ten men per tent. Our mattresses were stuffed with straw, and on them we found sheets and blankets.

Reveille was at 6:30 in the morning, and we began our day with putting our tents in order. A half-hour later we had our morning calisthenics. Then, shirtless, and with towels in hand, we would be divided into squads and run to the Oka River for our morning ablutions. The mass of half naked people lining the banks of the river was quite a unique spectacle. It is not surprising that those who weren't able to get close to the upstream part of the river had the worst situation - the river's current carried the soapy water downstream, and those poor souls had to wash in it. But you could not really think about this because in a very few minutes we had to form up again in squads and run back to the camp's mess hall for breakfast, which began at precisely at eight in the morning. You had a half hour to eat before we were formed up into platoons and marched out to the training area.

Upon arrival at the training area we again stood muster, and sent off for our training. The morning was spent learning weapons, drilling and instruction in basic military matters. Our midday meal was at 12:30, and we were allowed an hour to rest before we were marched back to the training ground.

I spent the afternoons learning my specialty as a signaler. We signalers were formed into small groups and began learning the Morse alphabet in Russian from the very first day. We drilled and drilled until we could look upon the Morse alphabet as an art form. We were also instructed on using various radio and telephone equipment, along with simple encoding. I applied myself to my craft, and within a short time I could tap out 80 characters a minute.

Our sergeant-major was a man named Mikhail, and was an old soldier who began his career as under the Tsar. He was a patient and excellent instructor who knew his craft well, and was able to pass his knowledge on to us. But there was something most unusual about our sergeant-major, Mikhail liked and respected us "Polish-Westerners," and during our free times, especially after he had a few beers under his belt, Mikhail would tell us colorful stories about life before the First World War and the Russian Revolution.

I still remember one of his stories. Until the First World War the Russians had numerous garrisons on Polish soil, which was then ruled by the Tsar. Mikhail was stationed there during his first years in the army, and told us about how beautiful the Polish girls were, and wistfully told us that he had fallen in love with one of them. But what came of this? She later fell in love with a Polish boy, and she was finished with Mikhail. Despite his great disappointment, the old soldier would share his memories of the Polish women, and how chic they were, and how they always felt a great deal of pride of not only their appearance, but also in themselves.

When the war broke out, there were many transfers to and from our division, and Sergeant-Major Mikhail was sent to another unit. It was a great shame in that he liked us Poles very much, and we knew that we would miss his solid personality, and the fatherly way he treated us recruits, and of course, his picturesque stories.

I am cannot over-emphasize the prominence that the "*politruk*," or political officer, had at all levels of the Red Army. Just as the company commander was responsible for training his unit, the *politruk* had a key role and was responsible for ideological instruction and for maintaining the morale of the Soviet soldier. In addition to this role, the *politruk* has the last word in carrying out military orders, and would overrule his unit commander's instructions on political grounds. This system was so widespread, that there were political personnel in even the smallest military unit.

In our section we had a junior corporal whose function was to report the conditions, morale, and state of the soldiers to the *politruk*. The one subject that the political officers never seemed to stop harping about was the "wonderful friendship" that now existed between Communism and National Socialism. This importance of this "monumental step in the progress of mankind" was repeated to us over and over again. The *politruks* paid special attention to us "Westerners," and we were constantly subjected to lectures about the superiority of the Soviet social and economic system. But we who had lived a life unfettered by ideology instinctively knew different, and had to keep a straight face while we listened to the nauseating rubbish that they repeated constantly, and it seared our very souls when we had to nod in agreement, or applaud their visions of the world. In later years, the continuous repetition of lies until they became the truth would be known as "brainwashing."

We were allowed free time to ourselves only after our evening meal, and this lasted until lights out. It was our opportunity to write letters home, and I again was able to communicate with my parents in Dubno. I also was able to establish contact with my sister and her family who had earlier been deported north to the Arctic Circle in the Archangel region. But as events unfolded, we exchanged only a few letters before would again loose contact.

During our free time the Lewicki brothers and I strolled through the entire camp, where we might get to know people, and more of this entirely new world we were now part of. In other units, I met many Poles were completely Russified. They spoke of having Polish parents or grandparents who had lived in Ukraine. One of them was a sympathetic and very obliging man named Vladia Kowalski. He was bright, but had been seduced by communist propaganda. Despite the fact that he looked both emaciated and frightened, we regularly met during our free time. It became obvious that he began to feel very comfortable in my presence, and began to trust and confide in me, although what he knew about Poland was totally distorted. Vladia's parents had come from Zhitomir, which always had a large Polish population, but they had been deported to Kazakhstan in 1936.

Despite their best efforts, Poles living in the Soviet Union refused to accept either atheism

Vladia Kowalski

or Marxist socialism, and great numbers were deported under conditions that resulted in many deaths. This also was the fate of many Polish families living in the Soviet Union living in the Belarus and Ukraine as the Soviets feared having Poles living anywhere near the border in case of war. Our pleasant friendship did not last long, because when the war broke out in 1941 we lost touch with each other, and I never saw Vladia Kowalski again.

I also became very friendly with a Russian in our camp. It was a young Soviet soldier named "Vovka." He was intelligent and very well read. Vovka had been a teacher before he had been conscripted. We are able to discuss many different topics, although he had been typically steeped in Marxist propaganda, he did try to see the world in a different light despite the fact that he had never been outside of the Soviet Union.

Our friendship was another victim of the outbreak of war, and the drastic reorganizations in our regiment. It was probable that Vovka had been sent to the front straightaway and disappeared into the vortex of war.

The month was coming to an end, and the days were becoming warmer. The everyday training routines of recruit training annoyed me less than they had at the beginning. After a hard day of training I was usually as hungry as a wolf, and fortunately, there was often something to buy in the canteen. I remember that halvah and cologne water were always available, and there were always soldiers who did not look upon the small bottles with the yellow liquid as fragrance, but as a "beverage."

My communications training was actually starting to become interesting. After I had learned the Morse alphabet, we began training with radio receivers. In a short time, I was spending my free time trying to tune in music from stations either in Moscow or Orel. But telephone communications did not have nearly the same interest for me. Learning the equipment was not a problem, but laying the telephone wire in a supposed combat situation was. I hated having to run like a maniac with a full drum of telephone wire balanced and having to lay out the line, then connect the telephone, establish

Vovka

communications with the telephone central, and then and after only a few minutes, having to go break everything down, and again start running with the cable reel to a new position.

It was very warm during the beginning of June. One pleasant Sunday afternoon a group of us left our camp and went for a walk along the Oka River. We decided that we would have a swim in the river. We hopped into the water. After a splashing around near the river bank, and verifying that everyone could swim, we decided to test our strength and race to the opposite shore and back.

No one among us wanted to display either fear, or admit being a poor swimmer. The swim across the river was no problem, but somehow the return trip was much more difficult. I managed to be the first to get back to our starting point, and from there, I looked back at the others. I saw that one of my friends was far behind the others. Then, his head disappeared. His head popped up again, but it was obvious that he needed help. I turned around and began swimming toward the struggling soldier. As I was getting closer I saw that it was Witold, and he was waving his arms rapidly, and was in a considerable panic. I managed to grab him by his leg. We struggled, but finally locked an arm around him so I could keep his head above water and began the task of pulling him to shore. I swallowed some water myself, and began to think that I would not make it myself, but in the end I eventually felt the river bottom under my feet. We had almost payed for our Sunday adventure with our lives. During our subsequent wanderings and perils, and right until the moment we departed the Soviet Union, I was blessed by having Witek as a friend, and he will always have my gratitude.

War
June 22, 1941

Then came that memorable Sunday, June 22, 1941. We were at the training grounds as usual, and I was monitoring the radio. During a break from duty, I turned to another frequency to see if I could get some music, but then I heard something I never before heard on a military radio set, a strange, emphatic voice speaking in a foreign language. Then it struck me, it was German. For a moment, I was disoriented, but quickly began to feel that something crucial was happening. I gave the headset to a friend, and his face became very serious. Neither of us knew German, and we did not know that the "Great Friendship" between Hitler and Stalin had burst, as if it was a soap bubble, and now we heard Hitler railing about how Germany would now bury Stalin and his Soviet Union.

We did not have to wait long for a reaction. At noon the entire garrison was assembled, and the chief political officer stepped onto an improvised speaking platform and roared through a megaphone that; "This morning the Fascist Germans attacked our 'peace-loving Soviet Union,' but we, under the leadership of our great commander, Stalin, we will annihilate the 'Hitlerite beast' forever."

I remembered when the Red Army had marched into Poland in 1939, and saw their soldiers, poorly uniformed and marching on in shabby, worn out boots with their rifles slung from their shoulders by pieces of rope. I did not think they had a chance to defeat the Germans. The *politruk* continued to ramble on, yelling that the fascist Germans (who hours before had been his "faithful friends," had unleashed brutal air attacks against Soviet cities and their helpless civilians, and whose tanks and infantry had crossed the border in many places. He then began boasting that the Red Army, "under banners of Lenin and Stalin would go forward to victory!"

And so began something that would have a very important effect on changing the fate of countless Poles who were torn from their native soil and sent to the remotest regions of the Soviet Union. The following day the entire regiment was mobilized, and throughout the camp there were frantic preparations. We were soon ready to return to Livny, and our barracks.

But what was happening? Nobody had any idea. They issued us new weapons and gave us a chance to fire a few rounds from them on a nearby range. The *politruk*s were now really over the top. At daily meetings they trumpeted how bravely our Red Army comrades at the front were flogging the Germans. We had to sit there and swallow this tripe because we had no other source of information.

Then one morning we were loaded onto freight cars in full combat gear. The question was were we in fact capable soldiers due to our short period of training? But at any rate, we were going to the front to fight the Fascists.

I cannot remember how many days we traveled, but there was little conversation, and our sense of foreboding managed to almost become greater than the need to put something in our rumbling stomachs. We had traveled almost 500 kilometers and were approaching the front in Belarus when the sound of the engine and the click-clack of the rails was drowned out by the same whining sound of German aircraft engines that were unforgettable after 1939. Then, with sound of machine guns and bomb explosions, everything went crazy as our wagon left the rails.

We were all a pile of arms, legs, weapons and equipment in one corner of the freight wagon. The sound of guns and bombs was replaced by the sounds of men screaming in pain. We untangled ourselves, and other than bumps, bruises and minor cuts, we were all in good shape. When I jumped down from the door of our wagon, the sight I saw was incredible. Streams of steam were spouting out of the engine, and many of the freight cars were smashed by bombs, or collision, or were on fire. Blood and body parts were scattered along side the track. We helped the injured as we could, but the entire scene was one of pandemonium. Many of the new Red Army soldiers stood around in shock. They were young, and never had experienced anything like this.

The losses were heavy, both in men and equipment, and our regiment was hopelessly disorganized. In such a situation it was unthinkable that our regiment could face an enemy in battle, so we went back to our garrison in Livny. We returned to a camp which now had a totally different atmosphere than it had been less than two weeks before. The previous order and discipline had given way to an atmosphere disbelief, disorder and fear. To our astonishment, instead of reorganization and return to the front, we were told that we were going to be sent back to our summer camp near Orel, and this time by rail.

The city of Orel had been bombed several times, and this accentuated the disorganization and paralyzed both the local citizens and what was left of our regiment. There we were told that the 400th Rifle regiment was to become a reserve unit, and would remain stationed in a training area. We "Westerners" who came from the *Kresy* acted much differently to the German successes than did the Soviet masses. They were deceived by a lifetime of propaganda and had been convinced of the invincibility of their nation, and now their faith began to crumble. We Poles, who had lived through the war in 1939, and most of us had seen the horrors meted out by the Luftwaffe on columns of refugees and towns and villages with no military value. But the Russians were now overwhelmed by fear and panic. They could not believe that their "best ally"was now flogging them mercilessly... wonder how this was even possible?

Events unfolded at a lightning tempo, which was matched by how quickly the front lines were crumbling. The camp and its order and discipline had radically changed, and we were now on a wartime footing. What we were fed had deteriorated to such a point that during training we were often hungry, and could not think about anything but what we could eat. During their daily harangues the *politruks* told us that the "Heroic Red Army" was withdrawing to better positions, which left us "Westerners" incredulous. In fact, we later found out that Soviet soldiers were surrendering to the Germans by the hundreds of thousands. The Red Army commanders found it hard to swallow that these mass surrenders were mostly by units composed on native Russians, and they now asked themselves what could they expect from Poles, Czechs Ukrainians or Belorussians who were forcibly conscripted into the Red Army? And, it was true, all we wanted to do was to return home, if the Germans would let us.

If the Germans had not bombarded our transport on its way to the front. I probably would have also submitted myself to do what I had to do. It was difficult. We were caught between the hammer and the anvil, and I personally decided that I would have no other choice but to fight. In the face of the deteriorating situation at the front, conditions became worse with each passing day. The Soviet leadership could not trust their own soldiers, had sworn an oath to uphold and remain faithful to Stalin's Soviet Constitution, and they were crumbling like sandstone. Needless to say, they especially distrusted us, conscripts who had come from "bourgeoisie Poland."

Then one day there was an early morning assembly and our *politruk* appeared together with the camp commander. They had a prepared list with the names of all the "westerners," who were considered an unreliable element. We were to turn in our weapons and uniforms. Forcibly conscripted into the Red Army from the eastern regions of the Polish Republic, the Soviets now feared that this diverse element may or may not have been relied upon to carry out orders. We mostly remained together in our own national groups. Ukrainians and Poles were in the majority, but there are also Czechs, Jews and Belorussians among us, and something had obviously been prepared for us, but what? None of us had the slightest idea what any of this meant. It was impossible to determine at that moment whether this was some sort of charade, or were we at the brink of some dire danger? Like it or not, we were part of a nation at war, and it was easy to imagine anything.

Events unfolded rapidly. Now degraded from the "honored and respected" status of a "Red Army man," we were now dressed in worn out uniforms, which were older than any of us. They still reeked of the Tsarist soldiers who had last worn them and were closer to being rags rather than clothing. We now had no idea what plans the Soviets had for us. All we knew was that we were loaded onto railroad freight cars on July 24, 1941, and started moving eastwards. I became deeply resigned to whatever fate awaited me. From my perspective everything looked entirely black. This new situation also meant that I would lose any and all contact with my family. Even Stanislaw Lewicki, who was always full of hope and enthusiasm that we would surely sometime return to our homes, shared my gloomy premonitions.

I was unable to find anybody who could tell us where they were taking us. The train was guarded by Red Army soldiers, and I attempted to get what information I could from them. But they themselves apparently did not know where they were going, or else they had been told not to say anything to us. The monotone click-clack of the train accompanied us on our journey east. We passed the Kazionka and Grazin stations, but after a short stop at the main railroad station in Penza, the train began traveling in a northerly direction. As it was the end of July, the days were warm, and we passed the time watching the landscape flash by through the open freight car door. The terrain was that of washboard hills planted with young trees, with the occasional *kolkhoz*, but overall, the landscape was fairly boring, and was nowhere nearly as attractive as our beautiful Volhynia.

As the days went by we began to feel hunger. They fed us once a day, the typical ration of black bread and dried, salted fish. People began discovering that their bread was being stolen, and eventually several thieves were caught. To our shame we discovered that they were from Polesie, the isolated swampy region immediately north of Wolyn. We ostracized them, and the thievery ceased. I was never able to explain to myself why this had to happen, and that the perpetrators were from Polesie,

With meals consisting of dry, whole-grain bread, without any lard or butter, and the usual salted fish, one craved water after eating. The thirst was unbelievable. We prayed and longed for the train to stop near some station. There one would find *"kipyatok,"* something that was known traditionally by anyone who was transported across the immense expanse of Russia. At every station along every stretch of railway, sat a small shack beside the main building where someone tended a fire day and night, which kept the water in the typical samovar close to a boil. When a train stopped a line would immediately form. Travelers would use the *kipyatok* to brew tea, which was laughable in our circumstances, and we, like the majority of Russians, would both slake our thirst and warm up by drinking the plain boiled water. At the larger stations there was even an opportunity to get a little soup served from a kettle. The "soup," for the love of God, was soup in name only, but we were able to faintly discern that there was something added to the water, and were very grateful for anything.

On the Road to a Siberian Labor Camp
August 5, 1941

Despite the fact that we did not have a map, we really did not have any difficulty in orienting ourselves in which direction we were headed. We often had to stop on the sidings near small railroad stations to give priority to other trains. At times we even managed to get some information. You began by asking, "How far has the Red Army thrown back the fascist invaders since yesterday? In such a manner we eventually found out that we were heading toward the Urals.

Between Penza and Kirov I was struck by the fact that the people building a parallel rail line to the one we were on, that the majority of their laborers were not Russian. They were short, had slanted eyes and a swarthy complexion. Who were these people, I wondered to myself?

My attempts to draw these people into conversation were futile. The most I could get out of them was muttering, and it was obvious that they did not want to have anything to do with us. But their faces with betrayed the fact that they were starving, and thinking that we were soldiers, they put out their hands begging for a piece of bread. Unfortunately, we had none to spare.

It turned out that these were Tatars, descendants of the race of warrior-horsemen who had overrun Poland during the 12th and 13th Centuries. There was no one in Poland who had not learned in school about the Polish knights who fought them in 1241, and of the death in battle of King Henry the Pious. We were told more about the Tatars, who were not encountered very often in this part of the Soviet Union, by some Russians. After the demise of the Golden Horde Khanate, the majority of them settled in the region of the lower Volga and Crimea. This was their situation until the Russian Revolution. The new government established the Tatar Republic, one of the many "republics" generated by the Soviet regime.

The Tatars had always been enemies of the Russians, and had fought them for centuries. I remembered reading how they had once sacked Moscow. But eventually they succumbed to the Tsar's armies. During the 1920s, they had resisted collectivization. They were deported to the depths of Russia In a very inhuman manner. As a result many of them died during the winters in the Kazan region where many had been deported. They were unable to deal with the change from the comfortable Mediterranean climate in Crimea to the brutal Russian winters. The men we saw laying the railroad line were some of those who had survived.

Genocide and mass relocation of minority groups were among of the specialties of Stalin's regime. The Tatars shared the fate suffered by ten million Ukrainians during the 1930s, who were murdered or died of hunger as they tried to resist collectivization of agriculture, and the resulting famine. We were always meeting Ukrainians who had survived in Siberia.

Kazan is a large, and ancient city on the Volga River, which is very wide at that point. We crossed the river on a long steel railroad bridge which had ships passing below. After another six hours, we crossed another river, the Vyatka. After another five hours, we crossed the Kama River. Both of these were tributaries of the "Mother Volga."

The landscape started to change as the train moved through the foothills of the Ural Mountains, which on the map were the dividing line between Europe and Asia. Early during the night of July 29, we went through the first of six railroad tunnels under the Urals, the longest of which was three kilometers long.

Though there was little to see in the dark, none of us could sleep. The question "where are they taking us?" was running through our heads. With morning the train emerged from under the mountains and we were on the eastern side of the Urals, and now in Asia. The train chugged onwards toward to Sverdlovsk. During the afternoon it arrived in the main station of that large industrial city. We could see with our own eyes, that we in fact were now in Siberia. Despite the fact that it was mid-summer, we were greeted by cold, drizzling weather.

After the train sat for several hours in the rail yard, we started moving again. We moved eastward an entire day, and the following afternoon we found ourselves at a small station. We were told to get off the train and were marched to a few kilometers to a barracks. There we met many Poles from the city of Lwow, and its surroundings. We apparently would be sharing a common future, which did not look bright, and we steeled ourselves for the worst. The barracks were ancient, had leaky roofs, and were, in all likelihood, part of an old labor camp that was built before the revolution. We slept on bare boards. Our meals consisted of foul-smelling dried fish, 400 grams of heavy, whole-grain bread. Once a day they served some sort of watery slop, which was supposed to pass for soup.

We now were working in the forests cutting down trees under military supervision. This was supposed to be temporary, and according to rumors, they would soon have us constructing an airfield, but who really knew? I kept having the horrible thought that we would have to spend the winter there. With what they were feeding us, without warm clothing and in shoddy barracks, not many of us would survive. We were working in the forest and the mosquitos bit us mercilessly.

At least it was the end of July, and we did not have to deal with the cold. During our free time we would disappear, either alone or in small groups, and pick blueberries. Our hunger was overwhelming. , and dominated all other needs, plans or desires. Hunger would haunt us until the moment we crossed the border and finally left the Soviet Union.

To our great surprise, after a week in this secluded wilderness, we received orders to get back on the freight cars. Our worst fears were coming to pass - we were again going east, deeper into Siberia. But to our surprise the train went west, back toward Sverdlovsk. Did this mean that we might be returned to duty as soldiers and be sent back to the front?

When the train reached the main Sverdlovsk railroad station, the train went off on a siding and into the railroad yard. When the train stopped, we were gathered together by the commander of the transport who was in the company of an officer wearing a blue visor cap. We immediately knew that he was with the NKVD - the People's Commissariat for Internal Affairs. This did not bode well. We stoically around the pair when the secret police officer, with a broad smile on his face, announced that the Polish government in London had signed an agreement with the Soviet Union, Stalin's pact with Germany was null and void, and further he stated that the Soviet Union would restore the Polish territories annexed in 1939, and that the Polish flag had been raised over the Kremlin. Now, as allies, a Polish Army would be raised to "fight the German Fascist beasts, shoulder to shoulder with the 'heroic' Red Army."

We were immediately swept by a wild joyousness that touched each and every one of us. Maybe our joy was a bit premature. We were then told our vegetation in our current circumstances for several more months. However, this announce did kindle a spark of hope for a better tomorrow. Perhaps the spark might even burst into a fire of expectation that we would eventually leave Stalinist captivity, and that we could, and would again, serve our motherland. We spontaneously shouted a vigorous "Vivat" for General Sikorski. Sporadic voices cried out, "Long Live Stalin," but we shouted them down. There was no way those voices were Polish. They were most likely Belorussians, who, from the very beginning, had sponged up Soviet propaganda.

We now knew why our transport had returned to Sverdlovsk. But this gave birth to another question: where would they take us now, and what would they do to us? We hoped that we would soon be taken to some sort of base where the Polish army would be formed, but it was obvious that the Soviets would not be willing to give up a source of free labor.

Our hopes were that we would soon be taken to some sort of base where the Polish army would be formed, but it was obvious that the Soviets would not be willing to give up a source of free labor. But in the meantime to show good will, and to confirm that there now existed a new regimen, and that in a sense we have now become "comrades in arms" they conducted us to a local canteen for a "good Soviet breakfast" This consisted of horsemeat soup with a generous portion of buckwheat groats added, 400 grams of bread and" chai "with a lump of sugar for a "*prikuska*" (an old Russian custom where a sugar cube is held in the teeth while the tea is sipped, rather than dissolved in the tea itself.)

However, the Sverdlovsk NKVD had yet to receive specific orders concerning our transport, and we were soon ordered back into the freight cars and further travels. Our transport was organized into three battalions, of some 700-800 men each. Most of these consisted of Poles, and for these travels, we were considered a sub-category of the Red Army.

We passed the city of Molotov on the Kama River. At the time I did not realize the irony that the city had been named Perm since the time of Peter the Great, but it was changed in 1940 to "honor" the Soviet foreign minister who signed the pact with the Nazis that divided Poland. The train now began stopping and waiting very frequently, sometimes for hours. We had no idea why. During these long pauses near small stations we had the opportunity to knock on the doors of the local huts to try to get something to eat. But food was only available if you had rubles, or even better, if one still had something from Poland to offer in trade.

We were young men, and eager for human contact. We were interested in knowing other people, and young women in particular. We saw many of these Russian girls engaged in various work throughout the Soviet Union. We even saw them engaged in very strenuous labor on the railroad beds, even laying rails. It was obvious; the men were away at war, and women had to take their places. I was struck by the fact that most them were wearing "*lapti*." This was footwear that was woven from the stringy sub-bark from linden trees. In Poland such shoes were seen only in the poorest and most remote villages. It broke my heart looking at those girls whose lives and youth were squandered by their circumstances. At the same time, I knew that they were born and raised on *kolkhozs*, and breast-fed communist propaganda, and didn't have the faintest idea of the world beyond theirs.

More and more frequently we were seeing trains packed with civilians, all headed in one direction - east. During one of the longer pauses on a siding, I discovered that the passengers on these trains were refugees from the larger cities, such as Moscow and Leningrad. The regime feared that these large cities would fall into German hands, and that the evacuation of part of the civilian population had been organized earlier, primarily members of the Soviet elite, and classes. skilled industrial workers.

From time to time we were stopped to give priority to passenger trains loaded with wounded from the front. I wanted something to smoke very badly, and I thought that the wounded soldiers, in such a situation, would be well provided for. I was not wrong. I went over to the train, and struck up a conversation with the wounded Red Army soldiers. Under the circumstances, I was dishonest, and told them that we were on our way to the front, and on the way to Kirov, where we would be issued proper uniforms, and equipped with new weapons, to go out and get the "*Germantsy*." Now established as a comrade-in-arms, the wounded soldiers gave me several packages of cigarettes. They were real cigarettes, made from Turkish tobacco, and no doubt reserved for the privileged. I had hardly seen a factory-made cigarette since leaving Dubno, and like everyone else, smoked "*kurishka*," which was made from the from the veins and stalks from the tobacco plants, rolled in a piece of newspaper.

They wished me all the best luck, and spoke from their hard earned experience, and warned me that it was not that easy to fight against a well-equipped and well-trained Nazi army. They warned me there was a very good chance that I may have to give my life for "our Mother Russia."

The Road to Prosnica

We continued traveling toward the northeast. I learned from a railroad man that this line went to Kirov. It turned out that I had told the truth to the wounded soldiers without knowing it. But little was predicable in Russia during 1941, and we never reached Kirov. On the morning of August 6, 1941, the train stopped in the middle of nowhere, and we received an order to get off of the train with all of our possessions. What in the world did they intend to do with us in this wilderness? The now empty train began moving while we were formed into marching columns. The section leaders took roll call him meanwhile we were wondering what all of this meant. After all, in Prosnica, which consisted of a few dozen small wooden huts built along the railway. All you could see , as far as the horizon, were endless forests or rolling fields.

The roll call organization and preparations for our march took several hours. It was afternoon before the commanders of the different units announced that there would be a change in our status. We would have new commanders, half of whom would be civilians. We would continue to retain a military structure and would remain subject to military orders and discipline.

One of my friends, Adam Klimpel, could barely stand on his own two feet during this "inspection and review." He had symptoms of malaria - alternating chills and fever. We wondered how we could possibly help him. I felt bad for this intelligent young man from Lwow, as we had become friends and were always able to discuss many different subjects.

Just before we left this "Soviet paradise," not knowing that our paths would never cross again, this same man wrote me the following:

We did not know each other long, my new, dear brother.
Born in the same nation, but knew each other only in another.
We are connected by the same ideas and united by the same thoughts,
And that is why, my countryman, these words I do not write for naught,
To beg that you remember me how I was during the misery we shared
How we ate, slept, and our words kept us from despair

By evening some farm wagons, each drawn by a scrawny horse, arrived to transport our provisions, our indispensable equipment, and with what room that was left over, the ill among us. There were probably requisitioned from the local *kolkhoz*s by our "commanders," and we managed to find a spot in a wagon for Adam. In my notebook I wrote the following:

"We marched out of Prosnica to the camp in a long column. In the meantime we found out that this was the final stage of our wanderings. We arrived late at night. We found ourselves 9 kilometers from Prosnica."

On the edge of a forest clearing was a tent camp, and although they showed years of constant use, they would manage to provide a roof over our heads. During the next few days we pulled boughs and branches from the young trees and used them as make-shift beds so we would not have to sleep on the bare ground. All of this struck us as very primitive and temporary. But of course, as always, those in charge "solemnly promised" that building materials were on the way. We needed lumber desperately, but it was impossible to make trees into boards with an axe.

We were sleeping ten to a tent. It is not an exaggeration to say that we were living like gypsies, and just like them, we spent the evenings warming ourselves by the fire. At night we plundered the fields of neighboring *kolkhoz*s looking for peas and young potatoes. The potatoes were freshly baked in the fire, but our mouths still had an aftertaste after so many weeks of eating dry and salted fish.

We built our fires in the forest so we would not be caught stealing "state property" from the *kolkhoz*, and never took any of the baked potatoes back to our tents. Hunger is a horrible thing, especially to the young, who will try to get something to eat at any cost, and is extremely demoralizing. Our night forays to the *kolkhozs* become more and more difficult, but to escape from the camp was hopeless - where could we escape to?

Prosnica
August 15, 1941

Just as the railroad men told us, and so it happened - we were there to build an airfield. The swift German advance had gobbled up huge amounts of Soviet territory, along with airfields, which had fallen into their hands during the early days of the war.

The surveyors had completed their work. On the areas marked out for the airfields there were literally hundreds of one-horse farm wagons, requisitioned from the state and *kolkhozs* from all over the region. These wagons were driven primarily by women or men who were too old for the army.

Now we were to fulfill our role, accompanied by new misery. We were divided into small groups and equipped with shovels. We were to fill the wagons with earth, and then take it to the low areas of the "airfield" to fill them in. I ask the reader to imagine a hilly field several kilometers square, and the entire area had to be leveled by the gigantic effort of several hundred slave workers, who were deprived of basic machinery. This extremely difficult task had to be accomplished by poor, starved, worn-out remnants of humanity with hand tools.

With each day hope continued to fade that we would ever escape this cursed existence, and re-join our comrades in the ranks of the Polish Army. The days were long and monotonous, but most of all, exhausting.

Then an unusual event. On Sunday, August 17, we had a normal reveille. Everybody preparing to face another day of hard labor. I do not know how it started, but the word went like lightning throughout the tents that we would not go to work that day because we were hungry. In fact, nobody went. .

We wondered what the reaction would be. They promised that we would be severely punished. That night, three of our fellow countrymen deserted the camp, somehow harboring the hope that they might somehow get to the western allies. Among them was my friend Adam Klimpel.

Despite our constant hopes and dreams, there was really no way that escape would be successful. On their third night of freedom, the trio was arrested while trying to board a freight train. After they were brought back to the camp, they were separated and put into solitary confinement until the matter would be seen to. The *politruk* told us that they would be turned over to Soviet military justice to insure that any further escape attempts would not be tolerated. Two NKVD officers arrived from Kirov. From a specially constructed stage they convened a court martial which the entire camp was forced to watch, and would serve as a warning to others.

In the beginning we feared that our friends would be shot. But apparently some other factor was at work. Within the Soviet legal system, a mere accusation meant that obviously, the accused was guilty. In a universe where the sun, moon and all the planets orbited around Stalin, punishment was swift, and, as the millions of graves across the length and breadth of the Soviet empire testify, punishment was harsh. When the sentence was pronounced, we were all stunned by its leniency. Our fellow comrades were sentenced to loss of all "privileges," confinement to camp, and a diet of bread and water. They were also warned not to repeat such behavior in the future. Later we understood the reason for such a light punishment. The Sikorski-Mayski agreement had been signed to raise Polish military units to fight the Germans alongside the Red Army. From that time on, no one harbored any thoughts of trying to escape.

Stanislaw Lewicki continued to supply us with tobacco, but he was not always able to scrounge something from the wounded Red Army men passing through Prosnica. I had always known that Lewicki was both very clever and very energetic, but I never was able to figure our how, and under what circumstances, he was able to shake loose bits and pieces of information. At any rate, the camp authorities needed someone who would daily walk the ten kilometers to Prosnica to pick up newspapers and mail. Naturally, Lewicki managed to get this job. This was very fortunate for us, because he would share information that he had learned from the local Russians, or that he had read in the newspaper *Pravda*.

We were constantly bombarded with such phrases and slogans as "our heroic army continues its ongoing struggle with the Hitlerite hoards, inflicting bitter losses on the enemy." This message was repeated more-or-less daily in various versions. We had long ago learned how to interpret and "read between the lines" of what we were being told. We instinctively knew that the Red Army was actually in retreat along the entire length of the front, and we had heard of the mass surrenders that took place during the early days of the war.

Then, one day Stanislaw Lewicki came to us with some real information that he had read in *Pravda*, that a Polish military mission from London had arrived in Moscow from to discuss the formation of Polish Army units within the Soviet Union. He also shared the surprising news that several articles in that official newspaper spoke of the Poles, Poland and her soldiers in most glowing terms.

My mind speed forward to the probability that the agreement concerned all of the Poles deported to the Soviet Union would also apply to my sister Fela and her family as well. I was certain that she also awaited release from deportation in the Archangel-Kotlas region, and perhaps her husband would be able to join the newly established Polish Army in the USSR. I wondered whether they even knew about the signing of the agreement, and if so, how great was their joy must be. We could not wait to return to Dubno, but that was, of course, impossible at the moment. At any rate, I thought that at least I would no longer be a slave laborer knocking down trees in the frozen north.

Shortly after the Polish-Soviet Military Agreement was signed, all everyone talked about was joining the Polish Army. There were also rumors about the possibility that we would be evacuated south, and sent to Persia. Surprisingly, this speculation actually came true for the Polish soldiers and their families. In the meantime, I continued to worry about the fact that I was unable to contact my sister because of our frequent changes of location.

People were making plans, even we be unable to live under the conditions in which we found ourselves. Our minds and bodies were deteriorating and one had to maintain a strong inner spirit to remain unbroken. But we were approaching the end of our strength and tolerance.

One day, during work, I had a conversation with our Russian brigade leader. I learned that carpenters would be needed to construct some airfield buildings. I got in touch with Witold Lewicki, who I had known to be an excellent carpenter. I knew, that with the two of us working together, we would support each other. That did in fact happen. We were to report to the carpentry shop at the nearby collective farm where we would be preparing the building materials. From our group of friends there were three of us there. We were also joined by Emil Smoliński, a quiet and intelligent fellow from a village near Lwow.

The carpentry shop was in the village of Igorka, some two kilometers from the camp. It provided work for twenty people, mostly from our camp. We worked with a roof over our heads. It was a luxury to no longer be exposed to the weather. Most of all, it meant no more long hours of backbreaking work shoveling dirt into wagons. After a week of sawing boards by hand, I was then given a hammer and a chisel. I began to think that during this time I had learned everything that there was to know about the art of carpentry. As a result I energetically pounded grooves into the prepared pieces with my new tools. Poor Emil, who was working alongside me, warned me with a broad grin that I should work a bit more slowly or I might be given the *"Stakhanovite"* production award by Stalin himself.

I needed no urging, as my enthusiasm evaporated when the carpentry shop job was not what I had imagined when the chisel slipped, and my hand suffered a very nasty cut. I returned to the camp with my bloody hand all wrapped up. Natasha Ivanova, the sympathetic camp woman doctor, bandaged my wounded hand properly. I managed a few days off.

During this time there was nothing to interest me within the camp, so I killed time by sneaking into the woods to pick blueberries. It was the end of August and I found quiet glades abundant with the succulent fruit, which was an unimaginable contrast with the disgusting slop we were forced to live on. for months. I stuffed myself and picked a good amount for my friends who shared my tent. After they returned from work, I brought out the berries. Feasting on the berries most contentedly, my friends appreciated my thoughtfulness.

Despite all of the promises from the camp authorities, there was no change or improvement in our nourishment. The "menu" was always the same - heavy and hard wholewheat bread, watery soup (once in a while we were lucky enough to find a bit of horsemeat in it), and salted fish. Occasionally we were treated to a couple of cubes of sugar. But there was virtually no fat in our diet - typical prison fare. It was not surprising that many of us suffered from dysentery. We also lived in fear of diseases such as scurvy beriberi, or pellagra, due to the lack of vitamins. We had not seen any vegetables or fruit for the longest time, except for the blueberries I had found; and blueberry season was very short.

I still had not received any information from my sister's family, who were somewhere near the Arctic Circle in the Archangel Oblast. I had posted two letters to her from Prosnica, but often wondered if she was already on the way south? After all, anything was possible.

September 1, 1941

This day marked the second anniversary of the treacherous invasion of Poland by Nazi Germany. The villainous pact between Stalin and Hitler was in fact the Fourth Partition of Poland, this time without the participation of Austria. Thinking about this throughout the day, we felt only bitter reflection on the cruel fate that had befallen our nation, and of our short period of independence after the bloody struggles during 123 years of foreign domination.

Although we now prayed for the freedom we had lost, and longed for home, we fully realized that Poland's geographic position could not have been worse. I remembered the words of Hubert Olbromski, a character in Stefan Zeromski's 1912 novel, The Faithful River, set in Poland during the 1863 January Uprising and its aftermath. Olbromski was a wounded soldier being nursed back to health by a young woman in charge of the manor house of her absent owners. He told her: "The Polish nation has found itself between two mill stones, and itself must become a millstone, or it will be ground up and consumed by Germany or Russia."

A great part of our national tragedy had its beginnings in the

16th Century when the powerful Polish nation began to decline due to its own peoples' love of liberty. And this love of liberty led some of the most disgraceful chapters of our history. Power was wrung from the Polish Crown by the nobles, who composed a large and privileged section of the nation. They forged the infamous Liberum Veto, which entitled one representative to stop any important legislation in the Polish Diet. Russia and Prussia took advantage to influence sympathetic Polish nobles into frustrating any legislation that would have strengthened the nation, which was governed by law, and practiced religious and ethnic tolerance -then unknown in Europe. As a result, Poland weakened into a state powerless to deal with her neighbors, all of whom ironically used a black eagle as their national symbol.

September 8, 1941

After the camp doctor pronounced my hand to have healed enough to work, I returned to the carpentry shop.

The village of Igorka was the home of a *sovkhoz*, or a "Soviet state farm." The *sovkhoz* differed from the *kolkhoz* (or collective farm) in that the *kolkhoz* was composed of the peasant farmers and their lands being brought together into a "collective," where all would share in the work and the profits. The problem was that there were rarely any "profits," due to mismanagement. The majority of the agricultural produce was taken by the state. The only way that most anybody on a *kolkhoz* survived was through the what they produced on the small allotment of land granted to each member family. While serfdom might have been abolished during the 19th Century, the resident members of the collective farm were not permitted to leave, except for military service. discipline was maintained by punishment, usually with the family allotment being confiscated, or the "troublemaker" would be sentenced to hard labor in the Gulag.

The state farm was created from land confiscated from large estates, or virgin wilderness land. The labor force usually came from poor, landless peasants, who were paid employees. Just as on the *kolkhoz*, the employees of the *sovkhoz* farms were rarely given permission to leave.

The state farmer was managed by a government-appointed director, who developed the enterprise with state money. This was a major reason that rising stars in the communist administration sought to "build socialism" on a state farm, rather than a collective farm, as it was easier to get something to eat there. After all, everything revolved around an empty stomach.

There were other occasions when I met elderly collective farmers, and one of them told me: "Our people have learned not to work honestly. Everything in life is now dependent on robbery and fraud. Everyone is out to deceive each other, and one nobody has any faith in anybody else. The Director of the State Farm, the

Chairman of the Village Soviet, the bookkeeper, the auditor, the director of the Tractor Cooperative, five brigade leaders, the warehouse manager and his assistants - who could possibly maintain a proper accounting? Once, back when, people were all of a family, and everybody worked, and was properly paid, but them? They steal everything among themselves."

If we defeat the "*Germantsy*," we again "liberate" you, then everything will be the same as it is here in your Polish nation. Your people will also loose your honesty. None of you will work for so many "Lords." *

The day was Saturday, and as with any other day, Saturday was no different from any other working day, and we worked from dawn until dusk. Instead of the company of my friends during our half-hour lunch break, I took a walk to the to the tractor barn in Igorka. The majority of people who were working there were women, began a pleasant conversation, and began to get to know each other. I was so involved with Katya, that I even missed eating during our break! She told me that there would be dancing to accordion music during the evening. At her request, I promised her that not only would I show up, but that I would bring some of my friends.

Right after work, three of us went over to the village. I was joined by Emil Smolinski and Stanislaw Lewicki. Despite the fact that Lewicki all ready had a girlfriend in Prosnitsa, he was not bothered by the thought of exploring "new horizons" at the state farm. Since the affair was hosted by a Soviet state farm, the dance's theme was one of the usual, this time"All Youth Are at War." That was especially true for the local menfolk, almost all had all ready been conscripted by the Red Army. Many of them were all ready at the front. This left the young women to entertain themselves. They were really looked forward to the opportunity to be dancing with men, rather than with
 each other.One had to be in Siberia for a while, even a short while, among the Siberians to learn about their life and traditions. There is nothing more natural on the face of the earth than the simplicity and openness of the average Siberian. After twenty-odd years, the so-called leaders of the workers and peasants had not improved the living conditions of these people. The stigma and resentment remained. It was reflected in the witty but sardonic banter, and the sad songs which I would soon be hearing.

As we approached the barn where the dance was being held, we heard accordion music. Lonia played the accordion but all eyes were on Marusia, who was the center of the performance. She had a fantastic vocal range, and a beautiful voice which as the pure as the first dew-covered lilacs of spring.

I was riveted by the first song, about bandits who lived on the banks of the Amur River, which at its source, separated the Siberian far east from Manchuria. My heart took wings with the accordion's melody, while the lyrics but the words were both mirthful and very clever.

This was my introduction to the "*chastushka*." The *chastushka* is difficult to explain, in the same way that the Cockney rhyming slang is incomprehensible to a Londoner living a mile away in Mayfair. It is a characteristic Russian song, similar to a limerick, but with four lines of rhyming verse. The *chastushki* I heard that evening varied from sad and lonesome laments, professing love and longing, to others that were ironic or nothing short of obscene. They told stories about the young lassies waving goodbye to their soldier boyfriends, while anticipating the arrival of two new regiments to take their place... Of young lovers tenderly holding hands, on their way home from a dance, wishing that the road was longer, and they would not reach their destinations so soon.

Other songs enthusiastically soared with stories of Siberian bandits who enjoyed a freedom that these poor people could only imagine, having never been beyond the horizon from their state farm. The pace changed, and we heard the story of a poor peasant lad carrying a piglet, so in love with a girl who paid him no mind that the entire way he was kissing the little piggy while thinking of the girl.

The music went on into the night. The songs, whether warm and loving, sad or sarcastic, touched emotions that I thought had disappeared during that September of 1939. What all the *chastushki* had in common was that they were all funny beyond belief. The combination of the beautiful voices, the soaring accordion melodies and the raucous lyrics provoking the first sustained laughter and merriment I had experienced in what seemed an eternity.

Naturally, that evening gave us a respite to our wretched existence and distracted us from our bitter loneliness and longing for home. We danced to the waltz from "Adventures of Ivan" by Khachaturian, as people paired off for walks that lasted until late at night. The young Russian women, no less than other Slavic women, sought romance, and when they did, it was with their entire heart. I am not exaggerating when I say that in such a situation there were no barriers, and love made its own rules. Before, I had often wondered why Stanislaw Wyspianski, the Polish playwright, painter and poet, married a peasant girl. I found an answer that night.

I saw Katya whenever I could get away from the camp. As we became closer I learned that although she had little formal education, she did possess much innate intelligence. In the beginning our conversations were measured, but in time she began to trust me more and more. She went on to tell me of her life in Igorka and about the local village soviet, and how miserable life was in a colorless and rundown Siberian hamlet. She was very interested in what our lives were like in "the West." I was able to support my conversations through the few photos I still had.

When she saw photos of the neatly dressed students in their school uniforms, and even the average people in the street, who were obviously not suffering nor exploited in bourgeoisie Poland, the stories of which were constantly pushed into their heads, she was astonished. She took all of this in amazement, and with tears in her eyes.

September 16, 1941

At long last, I finally received a letter from my sister. It was dated August 2, 1941, which was very fast considering the wartime circumstances in a nation where everything is prioritized for its own needs. She wrote that they had received the necessary permissions, and were preparing to leave Kotlas to go south where the Polish Army was being formed, and added that her brother-in-law was hopeful that he would be accepted in their ranks. . She also told me that she was well, but that the first steps of their journey was by raft down the Dvina River, and had filled her with fear.

Finishing the letter, I returned to the realities of camp life. Night had not all ready fallen, but the temperature did too, after all, we were in northern Russia. And so the three of us, Stanislaw, Witold and myself would slink quietly into the woods. There, we would disappear into a thicket and light a fire and brew tea from roots or dried blueberries. One would always feel better with something to get warm with, which was easy with no lack of trees around us. We roasted "borrowed" potatoes in the ashes, sipped our tea and reminisced about "the good old days." Those potatoes tasted so good that it was as if that had come from Zomera, the renowned confectionary bakery in Dubno. With the gray reality of the morning ahead of us in a cold tent, the two hours spent by the fire with friends with pleasant conversation caused both those hours and those "good old days" before 1939 seem entirely too short.

We loathed the walk back to camp, knowing that we would be sleeping packed like sardines, curled up next to each other, and covered with our old and threadbare army overcoats in an attempt to stay warm. It was no longer a surprise to us that for over a week that our tents were covered with white frost every morning, a foretaste of what we would be facing shortly.

September 21, 1941

I must have been getting a cold because by evening I was not feeling myself. That night I was unable to sleep, I had the chills. Right after reveille I told Stanislaw that I was going to see the camp doctor because I had a temperature. To be released from work your temperature had to be at least 38.5 or 39 (101-102 Fahrenheit) degrees. Stanislaw told me that your temperature could be lower, and he actually had a point as he knew someone from one of the other tents had a thermometer, which naturally, had been stolen from a doctor. The man gave us a short lecture about how to get out of going to work. He told us how we could raise the temperature on a thermometer, saying that a thermometer held in the armpit would reach 39 degrees.

I went off to the doctor's office, and there were already several people on line waiting there to see her. The doctor put a thermometer in our armpits, and she stood there watched so no one could raise his temperature. When it was my turn, she calmly pulled out the thermometer one with 39 degrees, and I was released from work for several days. She then asked me with a laugh, how was my Russian? The question took me by surprise, and I replied that I was proficient enough in the Russian language to pass the examination for the certificate that qualified me as a teacher. She replied, *"Eta kharasho,"* (That is very good) and informed me that the camp commander needed a clerk.

I reported two days later, after all, I did have to get over my "Illness." I was made a clerk on the spot. The commander seemed to like everything about me, including my name. He asked if I was related to the author, Nikolai Ostrovsky, who wrote the novel How the Steel Was Tempered. The theme of the book was the Russian Civil War, and he was awarded the Order of Lenin, the highest decoration bestowed by the Soviet Union. I told him that we were not related, but within the very core of my soul I realized that had I been related to such a degenerate Stalinist toady, I would have preferred not to know.

With my new position, I thought I was in heaven. I was working indoors, and the office was heated! I was able to share my good fortune when the commandant gave me permission to get an old cast-iron stove from the warehouse for our tent. The joy in our tent was unbelievable, and the stove was installed without delay. We were all warmed just by the thought that we would not be huddled and shivering all night. We made a roster of who would wake up during the night to tend the fire. Fortunately, there was no lack of trees for fuel.

Among the most tormenting aspects of our life was that there was no way we might find out about what our situation, and how we would eventually join the Polish Army on the basis of the agreement. We so wanted to tear ourselves away from this wilderness, but that obviously would not happen anytime soon. What made all of this worse was that we could not foresee how much longer we were to be ignored and humiliated.

It was now the beginning of October, and the days were shorter and the weather colder. The sudden drop in the temperature fills us with dread. And though we have yet to have snow, there was frost everywhere by morning. The cold weather was all ready unbearable. We knew that unless our tattered clothing was supplemented, we would freeze. The food remained the same: slop soup, with a rare speck of horsemeat, a hunk of hard bread and salt fish.

We could see for ourselves that our so-called "construction project" to level the fields was nearing an end. But, we had no idea when the Prosnica airfield would be ready to have aircraft operating from it. Time was precious because the Germans had all ready seized quite a number of Soviet airfields during their advance. That was why everything was done in such a great hurry - the work was accompanied by a never ending chorus of "*Davai, Davai!*" (Faster, Faster!)

October 15, 1941

The time had come to take down our tents, roll them up and pack everything for the move. We sighed with relief that we would be leaving this boring place and its tents. The night were unbelievably cold, especially before dawn when our teeth chattered in the cold. We were to march to the Prosnica railroad station, but I began to regret the move because I feared that I would loose contact with my sister.

We were now on our way to our "new home," Nizhny Tagil, 127 kilometers southwest of Sverdlovsk. During the constant stops men would leave the train with mugs, pots or kettles to try to find some soup. Finding something to eat was rare, and they usually returned with the usual *kipyatok*, which did little to ease the pain of bowels twisted from hunger. We could barely look at the dried and salted fish, but we were given a piece once a day as our ration.

Whenever the train halted near one of the smaller stations, it was possible to knock on the doors of the track side cottages, and once in a while were able to exchange some of our meager possessions for some milk or a chunk of bread. However, I had traded my fountain pen and watch long ago, and had nothing else valuable left for barter.

During one of those halts, we had a pleasant surprise. There was another train also waiting, full of civilians. It turned out that they were Poles who had been deported to various parts of the Archangel Oblast. They had been freed according to the terms of the of the amnesty, and were heading south. I walked along the entire length of the train, hoping against hope that I would find my sister and her family. Unfortunately, they were not on this transport, but I did find two families from Bortnica, and another family from Dubno. They told me that my sister had all ready started on their way south earlier, and that it was doubtful that I would meet her.

The following day we went through Perm, and there, at the railroad station I met a one of my sister's friends, Mrs. Kaminska. I knew her only from my sister's letters, and when she found out who I was, was unable to tell me anything as her train already began moving out of the station.

Nizhny Tagil
October 16, 1941

We did find out that we were going to Nizny Tagil, a city east of the Urals. Our train again went through tunnels, retracing the route we had already traveled, but now, after finding ourselves on the eastern slope of the Urals, we turned south, rather than north toward Sverdlovsk.

. . . and so we managed to discover our final our destination. We later learned that Nizhny Tagil was selected to be an industrial zone near Sverdlovsk because of the excellent railroad network already in place there. Another reason for developing this area was in that it was east of the Ural Mountains, and the industrial complex would be secure in case of the outbreak of war. It should be noted that despite the Ribbentrop-Molotov nonaggression pact, the apparently the Soviets were counting on the fact that they would be drawn into war.

Nizhny Tagil had rapidly expanded into a gigantic industrial city. Many factories which were involved in armament production in the European parts of the Soviet Union were dismantled and then evacuated beyond the Urals in the fall of 1941, lest they be captured by the Germans.

The train carried us through the city, and deposited us in an open field. This was not a surprise to us, nor was the fact when they started dividing us into groups. "The commander" of our battalion told us what awaited us in the very near future - we were to spend the winter. Each group of some 40 to 50 people were to immediately begin work building their "*ziemlankas.*"

The word was unfamiliar to us, but we knew that it had something to do with they ground. After looking around the open fields, we noticed a wisp of smoke rising from the surface of the earth. In short order we learned that a *ziemlanka* was: a primitive shelter in the ground. First a large hole was dug, and then covered with branches and straw, or anything at hand, and then covered with earth. The smoke mysteriously coming from the ground was from a *ziemlanka* which was home to prisoners of war from Estonia and Latvia, captured when those nations had been annexed by the Soviet Union as part of the Ribbentrop-Molotov pact.

It was already late October, and here in the Ural Mountains there was already frost at night. We needed no urging to get to work on our shelters soon as possible. Everybody remembered stories from grandparents who had been deported to Siberia that the winters there were both horrible and merciless at the same time.

November 13, 1941

Our whole wretched existence during the coming winter months was to revolve around nothing but work and drudgery, the nearest thing to peace was when we returned to our *ziemlanka*. It was thirty meters long, ten meters wide, and two meters deep into the ground. The roof was supported by several posts which ran down the center of our new "home." Running along the entire length of the *ziemlanka*, a half meter from the dirt floor were our "bunks" where we were supposed to sleep. These were made from rough cut planks, full of splinters and knots. Of course, there were neither mattresses nor blankets, and we slept on the rough, bare boards. In the center of the *ziemlanka* was a iron stove, which was used for heat, and also to boil our *kipyatok*. As long as there was a fire in the stove, our *ziemlanka* was tolerable. However, someone had to watch the fire late into the night to keep it from going out. It was very bad during the early hours of the morning when the fire was reduced to embers, and we all had to huddle together for common warmth.

Reveille came at six o'clock in the morning. We stood in total darkness while the *politruk* took the role by the light of the lantern. He would then scratch their names off of his list of those who went to sleep the night before, and would never again wake up. We would then stand on line and had some watery slop made with horse meat or fish heads which they called "soup."

We were then given an 800 gram brick of hard and heavy whole-grain bread - this was supposed to last us all day. During the afternoon and night, we were again served the same so-called "soup."

We were unable to wash, where and how could we? So, it was not surprising, that suffering from malnutrition and a complete lack of any basic hygiene, the lice reigned over us, and multiplied at an unbelievable rate. We were taken for a steam bath once a week, and we turned over our clothing to be deloused in a steam chamber. The delousing steamer was not up to the task. The lice, exasperated and only "half-dead" from the experience, were even angrier after returning to their senses, and in a few hours, and again were burrowing into us eagerly. It was if they were getting drunk on our blood in order to forget about their time in the steam chamber. It was especially bad for us at night, when, instead of a well deserved rest, we had to struggle with this scourge, and often scratched ourselves bloody my morning.

In order to alleviate the misery caused by the lice we had to strip and stand naked and carry out a more effective intervention as we sought out the lice. Whoever did not do this would eventually have an open sore, and then an infection. Such intensive vermin hunting required precision skills: we had to crush the lice between

The lice usually stayed together in groups, but dispersed to carry on their "work" when a man was motionless or asleep. But, during the day, when one is active and in motion, they themselves lay still and rest where they were most comfortable, usually in the seams of our clothes and in the corners of our underwear.

Conditions in which we managed to survive the first days of winter in Nizhny Tagil foretold nothing good. Our lives again slipped into a vegetative state, and we languished and faced complete physical exhaustion of our very being. Our minds no longer were no longer sharp, and as a result, accidents became more frequent, and often with deadly results. We were so exhausted that people would go to sleep in the evening, and would not wake up in the morning.

My friend Wladek's face began to take on a strange appearance. He slept next to me in the *ziemlanka*, and I had noticed that his stomach was noticeably swollen. One morning, rather than waking up, Wladek just lay there dead. We later discovered that this was a consequence of eating filthy rubbish, which he tried to ward off his nagging hunger.

December 28, 1941

Just as in Prosnica, I again found work in the battalion's office as a clerk because of my knowledge of the Russian language. I became very dedicated this job so I would be able to keep it - I knew otherwise our fingernails, and not only the fully grown lice, but also the garlands of eggs that they had

My Polish Comrades in Siberia
Stanislaw Mazur   Kamil Smolinski
Standing
Stanislaw Lewicicki   Wladyslaw Lewicki
Adam Klimpel

laid. that there was either a pick of shovel waiting for me outside. The temperature was -40 degrees Celsius, and in such weather our battalion was preparing foundations for a future factory.

My job, consisted of keeping the register of who was in our battalion, and noting any transfers. I had to submit this daily to our battalion commander. The office was also housed in a *ziemlanka*,

In such a situation, under these conditions, when the matter became survival, I owed much to my friends, Stanislaw Mazur, and the Lewicki brothers. Stanislaw Mazur shared literally every morsel of bread with us. He was a "*Starszyna*" (literally "a Senior," - somewhat like a sergeant in the labor battalion) and always went wherever there a chance of getting something "on the side," or just simply stealing.

Just as in Prosnica, I again found work in the battalion's office as a clerk because of my knowledge of the Russian language. I became very dedicated this job so I would be able to keep it - I knew otherwise our fingernails, and not only the fully grown lice, but also

Winter 1941/42
I Had Removed All Red Army Insignia Before The Photograph Was Taken

the garlands of eggs that they had laid. that there was either a pick of shovel waiting for me outside. The temperature was -40 degrees Celsius, and in such weather our battalion was preparing foundations for a future factory.

My job, consisted of keeping the register of who was in our battalion, and noting any transfers. I had to submit this daily to our battalion commander. The office was also housed in a *ziemlanka*,

In such a situation, under these conditions, when the matter and that meant that I had to spend hours in very dim light. became survival, I owed much to my friends, Stanislaw Mazur, and the Lewicki brothers. Stanislaw Mazur shared literally every morsel of bread with us. He was a "*Starszyna*" (literally "a Senior," - somewhat like a sergeant in the labor battalion) and always went wherever there a chance of getting something "on the side," or just simply stealing.

During those very hard winter months of 1941-42, when Moscow was being threatened by the Germans from three sides. Those in the Kremlin saw their world slipping away from under their feet, and wanted to avert the approaching disaster at any price. So there were orders issued constantly to every corner of the Soviet empire to increase war production.

Meanwhile, the frost was constant, and the days dragged on. The gang leaders, under pressure from above urged the slave laborers; "*Davai, davai!*" They took the slightest pretext to discipline any evasion of work, and punished the offender by
by stopping his rations.

As my first Christmas away from Poland approached, I was miserable, being so far away from my family and friends. During the evening of our Christmas eve vigil, we gathered together in our *ziemlanka*. We were physically exhausted, but also filled with sadness and bitter reflection, and angry about our fate and our helplessness. We shared memories

As my first Christmas away from Poland approached, I was miserable, being so far away from my family and friends. During the evening of our Christmas eve vigil, we gathered together in our *ziemlanka*. We were physically exhausted, but also filled with sadness and bitter reflection, and angry about our fate and our helplessness. We shared memories of our homeland, and also of our carefree teenage years. Most of us tried, but we all failed to keep tears from forming. Crumbs of black bread replaced the "*Opłatek*" wafer that night beneath the surface of the frozen earth. The wafer was traditionally shared with loved ones during Christmas. The wish we shared along with the breadcrumbs with was the hope that we would get out of this hell as soon as possible and somehow return to normal life. We thought about the Polish Army being formed in southern Russia, and that the horrible war was raging everywhere.

February 25, 1942

Here, behind the Ural Mountains, far away from home and the front where a murderous life and death conflict was taking place, we remained under rigorous camp rules and discipline. The gang leaders and *politruk*s continued to fulfill the reason the camp's very existence, which was to squeeze the last gram of strength from man for the triumph of communism.

Despite our very difficult situation, we sought some sort of relaxation. By some great grace when received good news from the front of an imaginary counteroffensive against the Germans, which our *politruk* delivered with a broad smile. As a result, the Russians surprisingly treated us to a little relaxation. We were taken by tram the four kilometers to the city and taken to the movies.

I remember that the film was a propaganda piece titled "We Are From Kronstadt," lauding the heroism of the defenders of this fortress. History recorded that Kronstadt sailors, who had played an important role during the revolution, later went on strike, demanding "freedom of speech, a halt to deportations to labor camps, and an end to party control of the 'workers' councils.'" The secret police and the Red Army brutally suppressed the strike by killing many of the striking sailors, and deporting the rest.

Usually such movie theaters were filled with Soviet youth, again mostly girls, gnawing on the Russian national snack - sunflower seeds. The lucky girl who managed to attach herself to one of the few available males who was neither too young or too old was in luck. The frequent result was a fleeting love affair for the evening.

Meanwhile, the gray, and short winter days seemed to drag on forever. The freeze became very hard, and as the temperature plunged to -40 degrees Celsius, accidental frostbite was becoming more and more common.

Somehow during the middle of February we heard that the long-awaited representatives from General Sikorski's Polish Army in England were in camp to register recruits for the newly established Polish Army in the USSR. That day I tore myself away from my work streaked across the camp like a maniac to *ziemlanka* Number 27 to confirm that this rumor was in fact true. It is difficult to imagine my joy when I spotted two Polish officers! A young lieutenant of artillery was accompanied by an older man who wore the three stars of a captain on his shoulders and the cherry-red collar tabs of the Polish Army Medical Service. As always in the Soviet Union, a pair of NKVD men hovered nearby like guardian angels.

Registration began us during the evening - after all, we were supposed to work during the day. The commanders and the staff of the labor camp did little to conceal their annoyance with the registration. Naturally, they feared losing a slave workforce, and the difficulties in replacing the laborers during wartime. The NKVD also interfered with the recruiting commission's work. If anyone was judged not being purely Polish, their name was stricken from the registration list. As a result, Belorussians and Ukrainians who had been Polish citizens before 1939 were treated as Soviet citizens. Nevertheless, despite very penetrating investigations many of them managed to confuse and put the suspicious NKVD off guard, and in the end, often found themselves in the Polish Army.

The Polish recruiting commission managed to get the local Soviet authorities to agree that at the earliest instance, not longer than a week, that we would be able to leave the camp, and go by rail to Sverdlovsk (now known again by its former name, Yekaterinburg, and was the place where Tsar Nicholas II and his family were murdered in July, 1918). We were to appear before a commission there, where it would be decided which of us would be suitable for military service, and if accepted, they would receive railroad passes, and either as individuals, or in groups, would be moved south to the bases established for the Polish Army.

March 11, 1942

We left the ghastly factory city of Nizny Tagil, along with our fetid *ziemlanka*. Starving and in rags, and exhausted to the last sinews of our stamina, nevertheless we felt some hope that there were better days ahead. When we arrived in Sverdlovsk we received a very pleasant surprise: a large part of our group from Nizny Tagil was quartered in a recently closed musical conservatory! It was warm, and we received a bowl of soup, and a chunk of bread. We were beginning at last to appreciate some positive effects from the Polish-Soviet military agreement. The following morning we met with another Polish military commission, who further verified who and what we were. Later that same day they sent us to the railroad station to begin our journey south.

We came to realize first-hand the transportation problems in the Soviet Union. All of the trains that came through were limited to carrying soldiers or the produce of the military industries located east of the Urals. After 36 hours waiting for a train in the Sverdlovsk station, we eventually managed to get in an empty freight wagon heading south. Our group of twenty-five spread out and tried to make ourselves comfortable. We were most fortunate when we found out that it was equipped with an iron stove, even more fortunate that there were enough scraps of wood littering the wagon that we could make a fire. Even though it was mid-March, winter held western Siberia firmly in its grip.

Four days later, we arrived in Chelyabinsk, 195 kilometers from Sverdlovsk. We were hungrier than wolves, so with the Lewicki brothers, and two others we went to town to find something to eat. Other than a stopping at a barbershop, where a shave and a haircut gave us partial relief from looking like savages, we accomplished nothing.

After we had returned to our comrades in the freight wagon we began to think about doing something about our food problem. One of us sniffed out a freight wagon filled with sacks of flour on another track. Four of our group, just teenagers, stealthily went off, and how they managed I do not know, returned lugging a large sack. Mixing water with the flour, we made dough. We were in the process of cooking a crude noodle soup when suddenly we heard voices. It was a police patrol, and they were inspecting every freight wagon. It was not long before they were approaching ours, and we were in a panic about how and where we might hide such a large sack f flour. There was nothing we could do, and the four men who managed to "find" the flour for us were arrested, and taken away.

We knew that there was a Polish liaison office in Chelyabinsk, so we went there immediately to tell them what happened. We hoped, with the current "friendly" relations between the Polish and Soviet governments, that our liaison officers could some how intervene. There was not the slightest doubt, that the young men would have landed in the Gulag, sentenced to a number of years at hard labor for stealing flour earmarked to feed Red Army men at the front.

Our freight wagon, along with three others, were shunted off to a siding. You could not imagine my surprise when a train slowly passed, and we saw that the freight cars were decorated with red and white bunting and Polish eagles. This was a shock, as it was the first that I had seen any such patriotic manifestations of Poland displayed in public since that dark September over a year and a half before. The train came to a halt, and our cars were attached. The passengers on the decorated train where almost exclusively Polish, people who had been deported to the northern regions of the Soviet Union, and like us, were making their way south. Before the train resumed its journey I walked along the track, asking the new arrivals if they had any information about my sister, or if any one aboard was from the Dubno region. Again, no one was able to give me any information.

Because we were unable to take any real advantage of the sack of flour which was confiscated, our hunger remained. The following morning everyone who possessed travel orders received 600 grams of bread, and a piece of the usual salted fish. While we waited to leave, we were at least were able to get *kipyatok* to wash down the dry and salty fare. That was part of our reality that had not changed.

The most persistent and disturbing problem which confronted us was the fact that every time the train stopped, we had absolutely no idea how long the train would be there, and there was nobody who could tell us anything. Whenever someone had to get off the train to get something essential, such as water, we all too often lost somebody who had not returned before the train left. During our way south, we lost Stanislaw Tutaj in such a manner. The virtuoso violinist from Lwow had never left the wagon except for the one time when he went to get some water to wash with. The train started moving, and Tutaj was left behind. As the miserable little railroad station in the middle of the dull, snow covered landscape disappeared from sight behind us, we hoped that he would somehow catch up with our train. But we never saw or heard from him again.

March 18, 1942

After a nerve-wracking three days traveling, which played havoc with our morale, the train finally stopped at Chkalov. As the train came to a halt, we did not waste a moment. I, together with two others, raced through the station to the restaurant. It was always a case of first come, first served, and the first usually got the best of what was available. We were lucky, and managed to return to our wagon with twenty servings of soup and a bit of bread. Compared to our usual ration, this was a feast for our ever-empty stomachs which remained tormented and tyrannized by the salted fish.

The city had been named after a legendary Soviet aviator, Valery Chkalov. One of the local Russian women told us that before it had been called Orenburg (and it is again known by that name), and was an old, and historic fortress city at the confluence of the Ural and Ob rivers, and not far from the border with Kazakhstan.

Leaving Chkalov behind, the terrain began changing from forest to typical Asian steppe. Trees of any kind were becoming rare. Collecting wood for our stove became difficult. We began seeing fewer horses, and more donkeys and camels used as beasts of burden. Despite the fact that we were moving further south, winter still let us know that it was in charge, and the weather was bitter. The people who lived there swore that once we crossed the border into Kazakhstan the weather would be warmer.

Without any maps, it was very difficult to know where we were, or where we were going. The only way we had any idea where we were was when we would see a sign on a railroad station. Still, without a map even this information was little help.

We did know that our next stop would be Aktyubinsk in northern Kazakhstan. Our stop there turned out to be very short. We were told by some railroad workers that we were to the east of the inland Aral Sea. Our train would be going southeast following the northern banks of the Syr Darya River, following it all the way down to the city of Turkistan.

Kazakhstan was the second largest "republic" in the USSR Russia. It was also, after Siberia, became the second largest place of internal exile and deportation for "elements" who were either opposed to, or posed any threat to Tsarist Russia. This did not change after the Bolshevik revolution. Thousands of Ukrainians, Germans, Poles, and Chechens, were sent to this vast unpopulated semi-desert region. The rolling grassland had been used for centuries as the grazing grounds of the nomadic Kazakhs. Naturally, no one was permitted to leave.

March 22, 1942

The city of Turkistan in southern Kazakhstan had a very eastern appearance. It was on the on Silk Road from China. In fact, we were much closer to Chinese border than to Russia. Our stop again was very short. We left after we took on water. We were relieved not to see any more snow. The weather was much warmer. There was now one less thing to worry us - we would no longer fear freezing to death. Food was a little more plentiful here that in Siberia, primarily goat milk, and there was even butter available, providing you had something to barter for it. I had not had butter in so long, I had forgotten what it tasted like. We had little to barter, and at any rate, it was difficult to find bread. I did get to drink some goat milk, and will always remember its velvety texture and wonderful flavor.

After a dreary and monotonous journey of three days across Kazakhstan, the train arrived in Arys. Poles we had met along the way told us that our stop here would certainly be longer than those in either Aktyubinsk or Turkestan, and that our uniforms would be issued there.

Arys
March 25, 1942

Arys turned out to be a new town, less than fifty years old. Built around a major railroad junction, it was a major destination for Polish deportees from every corner of the USSR. As we wandered among the crowd of wretched people packing the market square, we heard many of them speaking Polish. They had arrived earlier, and like us, were trying to trade what few possessions we still had for something to eat.

There were no uniforms for us in Arys contrary to what we had been told. Today I am certain of the reason why. The attitude of the Soviet authorities sought to release the smallest possible number of Poles who were suitable for military service. This was happening at the same time that Moscow began to organize a Polish army which would be responsible only to the Kremlin to fight at the side of the Red Army.

But Arys will always remain in my memory. The town because it was there that I ran into three women from Bortnica who knew our family well. They had been in the same camp near Kotlas together with my sister. They assured me that most certainly she was all ready on her way south as her husband had all ready been accepted into the Polish army.

By the time the sun went down after our sixth day in Arys, we were exasperated having to wait in that dusty, horrible town. During that time, two full transport trains full of Poles had passed through the town on their way to Krasnovodsk on the Caspian Sea. Vodka was cheap, and many of the Poles stranded here drank themselves into a stupor. In our condition, it did not take much vodka to do so, and the apathy induced by alcohol seemed preferable to many than the nervous anxiety brought on by being so close, yet so far from our destination. They waiting for uniforms, proper food, and most important, information. I learned only later that the Polish Army was being formed in places other than Arys, such as Buzuluk in Russian, and Yangi Yul and the village of Guzar in Uzbekistan among other places.

My arrival to the Polish ranks were rather late in the scheme of things. In his book, THE INHUMAN LAND, written shortly after the war, Jozef Czapski wrote about the situation from the end of 1941:

"All through that month of March, 1942, our senior officers were fighting tooth and nail merely to keep our army on Soviet soil in being. The Polish high command was suddenly informed by the Soviets that the Polish army was to receive 40,000 rations as of March 20. Stalin during his negotiations with General Sikorski had agreed to raising an army of 92,000 men. At the beginning of March, we already had 75,000 men within our ranks, and this figure was going with every passing day. Despite the restrictions and the delaying tactics employed by the railroads, there was not a day when some new party of conscripts reached the Polish forces. This sudden reduction of our already insufficient rations provided the Polish Army was putting our soldiers into a potentially tragic situation.

"General Anders went to Moscow to see Stalin. During the meeting, which lasted only an hour and a half, Stalin demanded a reduction in the size of the Polish forces by almost 60%, and the remaining 40% were to go back to the collective farms, the mines, the forest and labor camps from which they came. General Anders knew well, if he agreed to Stalin's commands demand, that the result

would again be slavery and starvation for those soldiers. General Anders eventually managed to get the Soviet dictator to agree to evacuate those soldiers that the Soviets could not feed to Iran."

March 30, 1942

This day we left Ayrs, and our train pulled into Tashkent. This ancient city, one of the oldest in central Asia, was the capital of Uzbekistan. From a population of some 600,000 when the Germans invaded the Soviet Union, the city now held more than a million people. The Polish soldiers and civilians released by the Soviets were a only a minor factor in this population increase. When the factories were relocated to prevent them falling into German hands, tens of thousands of Russian and Ukrainian evacuees followed.

But we did not go to Tashkent as tourists, and we were soon on the train working our way through Uzbekistan. We eventually found ourselves in historic Samarkand, which was the capital of Tamerlane the Conqueror's empire. From the train we could see the city's past splendors, and the domes and minarets with their intricate mosaics in bright blues and turquoise, which offered quite a contrast to the sunbaked yellow countryside. Nobody dared to leave the train fearing being stranded after coming so far.

In short order we left Samarkand behind us. In our situation only one thing mattered, that we persevere in our efforts to get to the Polish Army. We were a bit surprised when the train started going west, and across the Amu Darya River, which formed the Uzbek border with Turkmenistan, another "republic" of the USSR. Turkmenistan was large country, and consisted of sparsely populated desert flat lands. The people who lived there were primarily shepherds, and lived in poverty.

It took two days for the train to cross the Kara-Kum Desert before we arrived at our destination, the Caspian port, Krasnovodsk (known today as Turkmenbashi). It was the morning of March 30, 1942 when we arrived at the dock. We saw what would be delivering us to Iran, an old steamship the you could barely make out "Baku," the ship's name on the rusty and flaking hull. The derelict should have been scrapped years before. But I did not hear any complaints as we moved up the gangway, and tried to find a corner while more and more people kept coming onboard. It was obvious that the Polish authorities were trying to get as many people out of the "Soviet paradise" before Stalin would again change his mind.

Krasnovodsk
Finally, Farewell To The "Inhuman Land"

Part 3

Pahlevi, Persia
April 4, 1942

In the end, we saw the "Promised Land" on the horizon - we had made to Persia. When we docked at Pahlevi, everybody left the ship in a hectic rush, as if to sever the final threads which connected us to that evil, inhuman land.

Here, on the Caspian sands of Pahlevi, we were swept up by an immense joyousness which words could never possibly express. This was the resurrection of people who were wretched shadows of their former selves, people who had literally returned from the grave and left behind the burdens, but not the memories, of deportation, hunger, exile, and the deaths of so many of those dearest to them. After one or even two years of misery and torment suffered in the various corners and crannies of the trackless expanses of Russia (very often in the footsteps of our ancestors), our appearance was similar to that of the millions of people who had been swallowed up by the Soviet Gulags. Here on Persian soil we presented quite a sight, that of a tragic mass of emaciated and ragged human beings close to the end of their tether. And that was what the local Persians and the small group of British military personnel saw.

Pahlevi - The First Shaky Steps To Freedom

The square at the end of our pier was dominated by a mountain of bundles surrounded by crowds of people with gray, gaunt faces. They were of all ages – adults, children, and the elderly - many of whom were so debilitated by typhus, malaria, scarlet fever, and other diseases epidemic in the USSR that they would not live long enough to appreciate their freedom. If one saw any smiles or other expressions of joy within this mass of numb humanity, it was because someone had finally found their loved ones. I was like the majority of the others, going through the crowd with the hope that I might possibly be able to find someone from Dubno, or others who had shared my experience while in the Soviet Union.

I was very impressed with the efficiency of the British soldiers who supervised the disembarkation of the ship and settling us into our new camp, just outside the port, where an entire tent city sprang up from the sand like mushrooms after the rain. It comprised over 2,000 tents provided by the Iranian Army. It stretched for several miles on either side of the lagoon: a vast complex of bathhouses, latrines, disinfection booths, laundries, sleeping quarters, bakeries, and a hospital. Every building in the city was requisitioned and every chair appropriated from local cinemas. Despite the efficient organization in place to cope with the task of more and more people arriving with every new transport from Russia, the facilities were still inadequate.

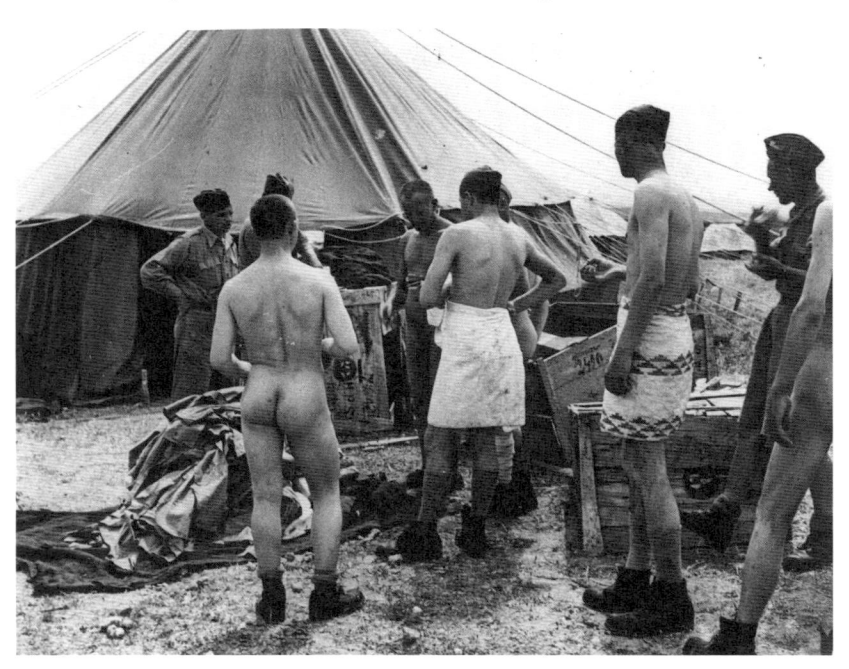

Out With The Old . . . .

Though the British were in charge, the camp was staffed by Persians, who provided all of the maintenance and labor. Never have I ever had the occasion to see anything managed so smoothly and competently organized. From that very moment, and through the rest of my years of military service, I admired how the British people managed to capably carry out whatever tasks they had with skill and proficiency.

On the second day we had to get rid of our old rags. That also meant everything we had, except for our most personal possessions. After we went through the disinfecting baths, we were given "haircuts," or more accurately, had our heads shaved. This was nothing unusual for anyone who had been conscripted into any nation's military. But the surprise came after they finished with our heads. We were ordered to keep our hands at our sides while our nether regions were also shaved. They were very determined that only people would be billeted in Pahlevi, and not any of the six-legged stowaways who had attached themselves to us in their attempt to breathe the free air of Persia.

We were naked as babies when we received new clothing. Those of us in the army were issued British so-called "battledress" uniforms. The others were given civilian clothing which had been provided by the Polish-American Relief Organization.

The large mound next to the camp continued to grow. It was composed of all kinds of crumbling suitcases, satchels, beaten-up trunks, wooden boxes and other items that held the pathetic possessions carried here by those whose good fortune permitted them to leave Soviet slavery behind. This heap soon had petrol poured all over it and ignited. Incinerated along with the items that people had carried over thousands of kilometers were millions of lice, who thankfully were no longer our persistent companions.

I felt refreshed and reinvigorated, and basked in the satisfaction at the ease and efficiency that my recent past was put behind me. My first night of freedom was chilly. We slept in tents, and by the time I awoke, I was hungry enough to eat a horse's neck, but I had to settle for a slice of South American corned beef from a can, and a mess tin of hot tea.

From the very early morning Persians were going down the lines of tents selling hard-boiled eggs. However, I had other things on my mind and did not forget for a second that above all, I had to find my sister. I did not have long to wait. My euphoria could not have possibly been greater when I saw her face. She and her daughter Irena had already spent two nights in Pahlevi, and by some sort of miracle, had arrived before I had. They looked gaunt and haggard, but all the same, they were relatively healthy. Little Irena, whom I had last seen in late 1939, shortly before they were all transported to the subarctic Archangel region, had managed to grow despite everything. Felicia told me that her husband had joined the Polish Army in Totskoye. He was again a sergeant, as he had been during the 1920 War, and had all ready been in Persia for several weeks.

Fela filled me in on what had happened from the time that our letters were interrupted, along with other things that happened to her and her family since they had been deported. She had much to say, including stories that could never be written in letters during Soviet captivity. Kotlas was almost a Polish village, populated to a large degree by Poles who had been deported there since the 1920s. The populations of entire Polish villages from Ukraine had been relocated during the Soviet war against the Kulaks.

The trek to reach the Polish Army was most difficult for those who began their journeys in the far north, as had my sister and brother-in-law. Not only had they further to travel, but they also had to wait patiently for the arctic winter to break. They were so far away from any sort of civilization that they had to build rafts, then wait for the Dvina River to thaw to get to the nearest railroad station.

My sister's journey from Kotlas after the amnesty was indeed frightening. The river was very swift. Not only was the rushing river dangerous because of the spring thaw, it was treacherous as a result of logging operations. The trees that had been felled over the course of the winter were trimmed, and then stacked on the ice of the frozen river. With the thaw, literally tens of thousands of logs went zooming downstream to the sawmills.

The raft carrying my sister and her family sped down the swift river, and all was fine for the first days of the journey. They spent the nights on shore, sleeping under the stars. One afternoon their raft was overrun by logs and suddenly flipped over. All of the passengers found themselves underwater, and any personal items which were not lashed down were lost to the river. It was a miracle that nobody lost their lives as they clung to the capsized raft as it spun in the churning river. It eventually hit a shoal, and the group managed to get it and themselves to the river bank. As they were salvaging anything they could from their raft, they were approached by a Russian peasant. He saw what had happened and invited them to get warm in his cabin on a bluff overlooking the river. The following morning they continued their journey. The difficulties and experiences were typical of the thousands of Poles working their way south and away from Soviet clutches.

April 4, 1942

I celebrated that most holy day, Easter Sunday, under canvas in a tent on the sands of Pahlevi. Instead of the bright colors of the traditionally decorated *Pisanki* eggs, we had to settle for the hard-boiled eggs purchased from the Persians. Rather than the Easter *babka* sweet bread, and ham with horseradish, we had mutton and dates. We especially appreciated the canned goods sent by our countrymen living in America who had collected money for the Polish American Relief Agency. But the fact was that we had survived. We were lucky, as were some of the others, who had found some of their family members despite the odds, and after all the toil, fear, hunger, and suffering we had experienced.

But now we were here in this paradise, a veritable land of plenty. However, our bodies were weakened by our experiences. After what we had suffered, it was very difficult for us to limit the amount of food, and especially fats, into our atrophied systems. Many of us were soon doubled over in pain, and dysentery and diarrhea swept through the camps. It was enough to eat a mere piece of sausage with the result of painful cramps and hours hovering near the latrines. There were cases of people who had experienced so much only to die a painful death by dehydration in Iran. I owe my survival to abstaining from food, despite my hunger. I kept repeating to myself: "Bolek, you went through so much deprivation and hunger, that you can also end up like those poor unfortunates, who ended their lives here, on the threshold of freedom."

The warm climate, healthy food, and above all, excellent medical care, had me feeling as if I was getting stronger with each passing day. It was not long, that after a number of examinations by army doctors, I was finally declared fit for military service. Soon, along with others, I found myself in a lorry on my way to Tehran. There, the British, with the permission of the Iranian government, maintained a large transit camp for the waves of Poles who continued to emerge from the USSR.

Pahlevi was essentially a reception area where the masses of former prisoners and deportees spent several weeks to regain their strength and health, and then were sent on to Tehran. My journey to Tehran was not much different than the others who experienced it. It could be summarized in four words: really horrible and frightening.

Our route went through the Elburz Mountain chain, whose highest mountain, Damavand, was 5,610 meters (18,406 feet) high. The young Persian lorry drivers were familiar with the route,

Polish "Tourists" Leaving Pahlevi

made dangerous by the narrow and poorly maintained roads that often ran between steep cliffs and yawning canyons of jagged rock.

We boarded the dependable Chevrolet lorries and began our climb up the narrow and winding road carved into the edge of the mountains. The drivers drove with a bravura that crossed the boundary into recklessness. As the lorry went through the hairpin turns, we were dizzy, and not only from the altitude. We did not want to see what was happening, but when we inadvertently glanced into the rugged chasms, one could see bits and pieces of smashed vehicles scattered among the crags below.

As much fear as the driving of the Persian maniacs caused in us soldiers, it was much worse on the women and children riding with us. It was almost a blessing that the wind roaring through the canvas helped to drown the screams of the women and crying children. When we finally made it across the mountains, it was with great relief when we were now on the straight and level road that crossed the arid plateau that led to the city of Tehran. Even so, the drivers seemed to forget that their lorries were equipped with brakes.

It was a great relief when we finally arrived in Tehran, but I am unable to say much about the Iranian capital city as we had no opportunity to see much there. We arrived on April 18, 1942, and were granted leave only once. But Tehran remains in my memory as a place of great contrasts - that of the wealthy elite who lived in oriental splendor, and the grinding poverty of streets filled with numerous beggars. But always, the Persians were friendly, and everybody greeted us with smiles.

In a short time, all of the Poles who had arrived in Pahlevi were brought to Tehran. Here, the Polish men; fathers, sons and husbands, were selected for service in the Polish armed forces, while their wives, mothers, and daughters were separated along with the elderly, or men unfit for service. They were to spend the war in refugee camps scattered among Africa, India, the Middle East, and even as far away as New Zealand or Mexico. We who had been recruited into the armed forces were sent to military camps, and we were handled in a military manner. Battalions were formed. We were assigned to our units.

We were then issued British battledress uniforms, which came in large bales, factory-fresh and smelling of moth repellent. I felt dashing in my new uniform, and like everybody else, adjusted my forage cap so it rested on my head in a cocky and confident gravity-defying manner. Now I really felt like a Polish soldier whose boots were planted on free soil. However, I also remained humble enough to thank God for keeping me alive to savor this moment.

My newly formed company, along with others from my battalion, was quartered in a large factory building just outside Tehran. Before the war, Germany had been supplying military equipment to Iran. Our factory billets confirmed this; the large production area was filled with tools and machinery for the manufacture of automobile parts.

After a two-week stay in Tehran, where our battalion's various sub-units were organized and their command structures staffed, we soldiers were then assigned to companies and platoons. All of those who signed up for service in the air force, the navy, and the recently created airborne forces were separated from the others to be transported to Britain. I found myself among those men as I, from the very beginning, had stated that I wanted to join the air force. The majority of the men would remain in the Middle East, where they would train and prepare for future operations in the Mediterranean region alongside other Allied armies.

All of these actions by the British military authorities were classified "Most Secret" to maintain security from German intelligence, which had made some inroads in the Middle East. As a precaution, among other counterintelligence actions, the Polish military units in the region were moved often. The training and educational facilities, along with staging areas for troops in transit, were secured by British soldiers. I was able to observe the efficient organization and supply of the various camps by our British allies, and was deeply impressed by the way they managed to make everything happen smoothly.

Our journey from Tehran south to Palestine began with the desert, and then ran across the rugged ridges of the Zagros Mountains on our way to Ahvaz. It was mid-April and the weather in southern Persia was already very hot.

I will never forget Ahvaz because of the sandstorms that plagued the region. I had never experienced a sandstorm. This was a totally new experience for me. It was like a dirty brown wall that stretched from horizon to horizon, and blotted out the sky, was coming toward me. I instinctively tried to stop breathing until I could get into a tent. But this was little help as the sand managed to get into everything. I had sand in my eyes, in my nose, and in my mouth, where it was crunching between my teeth.

The next stop on our way to Palestine was Baghdad. Our stay in Iraq was short, but fascinating. When we crossed the Tigris River I was astonished at what I saw. After all the countless kilometers of desert that we had traveled across since leaving Pahlevi, we now saw a green, lush land between the Tigris and Euphrates Rivers. I knew it as the place where Eve tempted Adam, called "Paradise" or "The Garden of Eden" by the priest who was our religious instructor when I was a boy. This green land surrounded by desert was, as was the Nile delta in Egypt, known to history as a "Cradle of Civilization," and supported life from the Persian Gulf to the mountainous desert near the source of both rivers near the Syrian border.

Again we were in lorries, driving through seemingly endless desert until we found ourselves in a Polish Army camp at

Al-Khassa, which was outside of Tel-Aviv, not very far from Jerusalem. Our tent camp was located over picturesque hills which proved to be excellent for military training. After a day of training or classroom instruction, we would feast on oranges, from the local Arabs who would come to our camp carrying the fruit on overloaded donkeys.

I never had any dealings with Arabs before, and I soon learned that not everything in the Holy Land was "holy." The nights were hot, and we slept with the walls of out tents rolled up so we might enjoy whatever breeze there was. The tents were round, and eight men would be assigned to a tent. We slept with our feet pointed to the center of the tent, and typically as soldiers, we slept with our heads on our knapsacks instead of pillows. We were warned that the Arabs were absolute masters in the art of thievery. They would be selling us oranges during the day, but at the same time they scrupulously noted what was around, and who had what. At night they would really go to work. Imagine our surprise and fury when we woke up one morning and discovered how easy we had made life for those who robbed us. They managed to pull the knapsacks from under our heads, which was not a surprise, considering that we were dead tired from a long hard day of training in the field. It was no wonder that we were sleeping like logs. In this way I had the few small items stolen from me that I had carried from Poland, including a religious medal that my mother gave me. From then on, whenever I was among Arabs, I always made sure that I was both careful and alert. This was a big change from what experiences we had in Iran with the ever-sympathetic Persians.

One very pleasant surprise about Palestine was that it became home to many Jews who had emigrated from Poland. We heard the Polish being everywhere. There were even Polish newspapers and radio programs. On weekends, there was time for sightseeing, either in groups or individually. We visited many of the places mentioned in both Testaments of the Bible. It was really wonderful seeing the marvels of the Holy Land, which I had known only from the Gospels, with my very own eyes, and to have my feet in the same places where Christ had walked.

After some three weeks in Al-Khassa, we packed up our gear so another group could occupy the area. We boarded a train which took us to Suez Harbor and the Suez Canal. There were even Polish newspapers and radio programs. On weekends, there was time for sightseeing, either in groups or individually. We visited many of the places mentioned in both Testaments of the Bible. It was really wonderful seeing the marvels of the Holy Land, which I had known only from the Gospels, with my very own eyes, and to have my feet in the same places where Christ had walked.

After some three weeks in Al-Khassa, we packed up our gear so another group could occupy the area. We boarded a train which took us to Suez Harbor and the Suez Canal. There were many other

Polish units with us. None of us had the foggiest idea where we were going and what route we would take – everything was a military secret. Shortly, we began boarding the RMS Mauritania. This was a large, pre-war luxury liner that had been converted into a troopship.

June 6, 1942

The ship cast off, entering the Red Sea from Suez. We then transited narrow straits into the Gulf of Aden, and then into the Indian Ocean. I was lost in thought, and like the others, wondered if we would be going around the tip of southern Africa and then on to Britain.

Around June 12, our ship crossed the Equator. The sailors, following tradition, managed to organize an initiation ceremony for those who had never crossed the Equator. We were inducted into the Kingdom of Neptune, the legendary god of the seas, and nobody, even officers, were spared the increasingly embarrassing ordeals involved.

The Mauritania maintained a southern course for the entire length of Africa's eastern coast. My comrades were unable to persuade me to take part in their card games. They were playing for money. Gambling never held an attraction for me. Instead I strolled the decks, exploring the ship's nooks and crannies, and enjoying my freedom. I enjoyed the idle tranquility and the fresh ocean breezes with the comforting monotone throb of the ship's engines as a background. Sometimes I would look in the direction of Africa, whose contours were barely visible on the horizon. Free from not having to deal with cold, hunger, and exhaustion, my mind began to fantasize that instead of being on a ship carrying me into the unknown, I was somewhere among the lush jungles of central Africa, lost just like the legendary Dr. David Livingstone.

Pietermartizburg, South Africa
June 16, 1942

After a week at sea, the ship entered Durban Harbor and docked. We wondered why we were not going straight to England. Near the pier a train with an electric locomotive was waiting for us. After we were all aboard, the train started moving west into the African continent.

After several hours traveling through very attractive green and hilly landscapes, the train stopped at a small but very neat railroad station. The signboard announced that we had reached Pietermaritzburg, and again someone was waiting for us. As we boarded buses we were amazed at the organization, and the sad contrast it presented to what our lives were like during the past years. The buses delivered us to a camp on a gently sloping hillside at the edge of town.

I later found out that this was a transit and rest camp for

Allied soldiers. Americans had been there before us. This was part of South Africa's contribution to the war effort.

I would like to think of our two-month sojourn in Pietermartizburg as one of my most pleasant memories, if it were not for the rather eccentric and despotic character of our commander. Major "M", was an elderly officer. I am happy to say that I do not remember his name. During our short stay he tried to make commandos out of us. As a result, the platoon leaders were urged to engage their soldiers in strenuous and challenging activities.

But the simple reality was that we did not even have any weapons to train with. We were given wooden sticks which were approximately as long and as thick as a rifle, which of course made us look and feel ridiculous. All the same, there were constant drills and parading, where we spent hours practicing the Polish Army goose step as if we were going to pass in review for General Sikorski himself.

As it had been only four months since we had left the "Soviet paradise," we were not in the best physical condition to be forced to participate in these demeaning and exhausting charades. We should have been fed and rested and exposed to light gymnastics after our experiences, so our bodies and minds might return to the physical and mental condition befitting a soldier before taking on such strenuous activities. Somehow, somewhere, someone should have changed the attitude of the egotistic camp commander who made us suffer for his whims, but it was not to be.

There was one very pleasant experience from Pietermartizburg that will always remain in my memory. The mayor organized a banquet as a gesture of friendship, so the Polish soldiers would remember their stay in that fine city. I was fortunate to have been chosen to be part of the delegation selected to dine with the municipal authorities during this happy occasion.

However, I was very sorry that I was unable to speak English at that time. I was seated between a pair of young ladies, but was unable to carry on any sort of conversation. All in all, I felt very bad because of this, and I promised myself that I would not spare myself in learning how to speak English. The stimulus behind this commitment was that there was no possibility of getting to know any of the young ladies from a Catholic convent school located not far from our camp. This was really a lost opportunity. It broke my heart that I was unable to express even the simplest thought in English.

Early in August 1942, we had to bid farewell to the hospitality of South Africa. Our company consisted of 100 young and rather inexperienced soldiers. None of us had the slightest idea what we would be doing during our further military service.

Upon returning to Durban's port, we were very surprised to find the RMS Mauritania again waiting for us. We now saw that there was something a little different about the ship. There were now machine guns positioned on the decks. We could not understand why they had these weapons, which were too light to be used as anti-aircraft guns, and why there was barbed wire blocking some of the passageways.

Not long after we had embarked we discovered why the ship had this unusual armament. We watched as over 2,000 prisoners of war, once members of Rommel's Afrika Korps, came on board. The Mauritania was to make a detour to deliver the prisoners to the United States before going on to our final destination in England. Our duty was to guard the prisoners.

The German prisoners of war kept below. They were under the command and administration of their own officers, who answered to the captain of the ship. I spent many hours under arms guarding them. My post was always the same, and I did manage to have contact with the prisoners on several occasions. I must say, that the German prisoners' conduct was exemplary. I admired them for the way they maintained their quarters and for their discipline, whether they were distributing their rations or carrying out orders. Despite the fact that the temperature below was very, very hot, especially when we were near the equator, they managed to carry on in those conditions without the slightest murmur.

Rio de Janeiro August 20, 1942

The Mauritania's destination was North America, but because food supplies were running low, along with the need for minor repairs, we entered Rio de Janeiro Harbor. As Brazil was technically neutral, permission was needed to anchor for a

RMS Mauritania, Luxury Liner Serving as Troop Transport

67

few days.

The view from the Mauritania's top deck was magnificent. The ship was anchored in the middle of a wide bay ringed by beautiful beaches with gold-colored sand. Towering over the city was a large mountain, surmounted by a gigantic statue of Christ the Redeemer. The statue, which was in the form of Jesus spreading his arms in blessing, was almost 40 meters tall. The sight of happy holiday-makers on the beaches was inconceivable to us newly-freed Soviet slave laborers, and filled us with envy. With night, the city's lights came on, sparkling like diamonds on the quiet water. The spectacular view was a startling contrast to the Europe where war was raging.

The following night a friend and I had guard duty on one of the upper decks. Long after midnight, we heard the sound of a splash, as if someone had fallen into the water. I looked over the rail, but in the dark I saw nothing except the twinkle of the distant city lights reflected in the water. A few minutes later, I heard a similar splash, and then another. I strained my eyes looking in the direction of the sound. My attention was now acute. Curiosity now had the upper hand over my guard duty obligations. What the hell could that have been? Suddenly, some 30 meters way from the ship, I recognized a head on the surface of the water. As I watched it disappeared, and then again came to the surface a short distance away. I did not have to think about the fact that this was an escape attempt, but brought my rifle to my shoulder and fired several rounds at the head. I then raised the alarm, but my shots had already gotten everybody's attention. Did my shot hit its target? I do not know, but now there were several more heads on the surface of the water. The other guards ran to the rail, and as soon as they realized what was happening, opened a furious rifle fusillade at the now surfaced Germans, who were frantically working their arms swimming toward shore. I could not imagine that anybody could have managed to swim the two to three kilometers to shore, but some of the escapees might have succeeded in reaching land.

As we later discovered, the Germans had access to a sink in the ship's galley where the mess cooks dumped kitchen waste directly into the sea. The drain's outlet was beneath the surface of the water, which caused the refuse to emerge some distance from the ship. This was the finale of a well-planned undertaking by the German prisoners. Rio de Janeiro was home to a large and well-organized German community who certainly would assisted the escapees in any way they could to have these men back in Wehrmacht uniform. During muster the following morning, it was very gratifying to hear the ship's captain praising us for our alertness and quick reaction in thwarting the escape attempt.

Weighing anchor in Rio de Janeiro, the Mauritania hugged the Brazilian coast as it moved north. The ship then threaded its way between the islands of the West Indies in order to reach the United

States as quickly and as safely as possible. German submarines continued to prowl the Atlantic waters and were a real nightmare for the Allies. They were the cause of enormous losses of men and tonnage, of both ships and their wartime cargo. The U-boats had seriously disrupted the shipping of vital materials to besieged Britain, and were a very serious threat.

The Mauritania, due to its speed and zig-zag maneuvering, managed to cross the ocean safely. Perhaps the enemy knew that the ship was carrying two thousand German soldiers. That was a possibility, but in the end we did not know if it was luck or the prisoners on board that shielded us. At any rate we did manage to get to our destination without any further incidents. Unfortunately, others did not share our good fortune , the RMS Laconia, another former luxury liner, was torpedoed with a heavy loss of life. Among those on board were 160 Polish soldiers guarding 1,800 Italian prisoners of war. They had left Durban only days after the Mauritania.

Newark, New Jersey
September 11, 1942

We arrived in the port of Newark, New Jersey, just south of New York City. With a great sense of relief, we got rid of our "ballast." The German prisoners of war left the ship and were now the responsibility of the Americans until the end of the war.

We later followed the German prisoners down the gangway, and with a certain amount of sentiment said goodbye to RMS Mauritania. While the food on board may not have been the best, we were grateful for a safe passage across the ocean. We were taken across the harbor to Fort Hamilton, in Brooklyn, where we would be billeted until it was our time to cross the ocean to Great Britain.

The commander of our detachment since we left South Africa was a law professor and reserve officer, Captain Bronisław Kusnierz. He would later serve the Polish government in London as Minister of Justice. I will always remember him as a likable, outgoing man who was very people-oriented. We owed him very much, and was in many ways responsible for our wonderful stay in New York.

The local Polish community pulled out all the stops when it came to hospitality. As soon as the news of our group's arrival spread through the city, there were literally dozens of automobiles waiting at Fort Hamilton's main gate every morning. Perhaps this was because we were possibly the first Polish soldiers in the United States recruited from those who had survived life in the "Soviet paradise," but whatever the reason, the way we were received was beyond description.

While growing up in Poland, I had heard and read much about that far-away America across the seas, but many things I learned seemed both contrary and controversial. Having witnessed the extreme poverty and dreariness of life in the Soviet Union, and then seeing the vigor and prosperity of the United States often left me confused. Even today, when I visit the land of George Washington, I still have to decipher what the actual situation is compared to my perceptions, which on occasion are still incorrect.

So, even in my newly issued British battledress, I somehow felt like a poor cousin in comparison with the ever-confident American soldiers. It was only much later that I was reassured and convinced myself that Americans in general have a very relaxed way of acting and doing things. I became a bit envious of them for that.

My early impressions of the American soldiers brought me back to my childhood memories of newly independent Poland. The majority of my colleagues came from the sleepy, isolated backwaters of the Polish borderlands and were usually poor, and often lacked the sophistication of those living in Poland's larger cities. As a result, many of my companions suffered the prejudices and primitive thinking that had become almost natural after the long years living under the Tsars.

The Americans were different from us in another aspect that was immediately visible: their growth rate. They were on the average taller than we were, which was probably a result of the abundance of food available to them during childhood. Our countrymen who had put down roots in the United States also differed from those of us born in Poland in this respect.

One evening we were taken out by a group of young Polish-Americans from New Jersey. Instead of bringing us to their homes, they took us all around New York City to see the sights. When evening fell, rather than returning to our billets at Fort Hamilton, the girls took us dancing at one of the local Polish clubs. As the night went on, one of the young ladies received more and more of my attention. She exerted such an impression on me that during the following few years she dominated my thoughts and the way I acted. That woman was Jozefa Kolwianka from New Jersey. When she spoke Polish, it was very proper and beautifully melodic, in sharp contrast to the butchered language that was typical of many of the Poles born in America. I later found out that Jozefa came to the United States with her parents only three years before the war broke out.

From that evening we met whenever possible. I no longer had any thoughts about sight-seeing in New York. I had seen everything I needed to see. Besides, I was a young man in love, love at first sight, confirmed by our first kiss. We were united by something that was much more than just friendship during that memorable time. It was a full-blown mutual love, but at the same time exacerbated by the tensions, anxieties and uncertainties of war. We promised each other that we would get married immediately after the war.

New York, New York

September 24, 1942

I was awake when the bugler blew reveille at Fort Hamilton, already packing my kitbag. Army lorries were to take us to the railroad station . At the main gate I saw a large crowd of Polish-Americans, waiting patiently. They were there to see us off. I scanned the crowd, looking for a familiar face, and especially that of Jozefa. . . and yes, I saw her on the other side of the street! I waved, she came over, and we held hands through the bars of the iron fence. As usual with hurried goodbyes, there was too little time even to speak in complete sentences. So, it was a last kiss and wishes for good health, a quick return, and pledges to remain faithful to our mutual promise.

Jozefa

Halifax Canada
September 29, 1942

We waited in Halifax several days while our convoy was forming. We had managed there from New York by train, traveling in very comfortable Pullman cars, and then embarked on the former luxury liner, SS Louis Pasteur, which after the fall of France had been used by the British as a troop ship, HMTS Pasteur. Halifax harbor was teeming with traffic, and we were surrounded by many transport and warships. This was an unusual and magnificent sight. I thought to myself that traveling in such company, I might manage get a peaceful night's sleep.

The convoy formed at the mouth of the harbor. It was a very complicated ballet where the transport ships carrying soldiers were in the center of the convoy, along with cargo ships loaded with priority items for Britain. In turn, those ships were encircled by freighters carrying lower priority cargos. The convoy was a mile wide, and stretching several miles in length. There were warships, including small aircraft carriers, while destroyers and other escort ships darted around and through the formation like shepherd dogs with a flock of sheep.

It must be remembered that the North Atlantic sea lanes were most dangerous areas, where the German submarines were most active. Although there were several U-boat alarms during our seven-day voyage, and dull explosions in the far distance, the convoy, as far as we knew, got through without loss.

Part 4

Greenock, Scotland
October 6, 1942

As I descended HMTS Pasteur's gangway I was smiling from ear to ear, a sign of the sense of relief I felt on again having the opportunity to have my feet trade the steel decks of a ship for the comfort of dry land. Despite the fact that we were sailing in convoy, defended by escort ships, there had been incidents were a German U-boat was not detected, found a gap in the convoy's defenses, and torpedoed transport ships. The fear that my life would end in the icy waters of the North Atlantic haunted me from the moment our ship cleared Halifax Harbor.

During those tragic days, Britain remained the beacon of freedom in the darkness of a world at war, and the hope of the peoples of Nazi-occupied Europe. Who from those generations does not remember the Morse Code "di-di-di-da" which was the signal for the letter V, which symbolized the coming Allied victory. These tones began every broadcast in every language beamed by the BBC before it broadcast its message of hope and liberation. All too often, the penalty for being caught listening to these forbidden broadcasts was death. Despite the fact that Hitler was in control of almost the entire European continent, he was unable to shut down this one pipeline of truth, and it drove him and his henchmen crazy. That was the atmosphere that reigned in Britain then, one of both faith and hope, where everything was directed toward the war effort.

Shortly after landing we were sent to the Polish mobilization center in Auchtertool, where hundreds of soldiers were gathered for assignment to the different Polish units stationed in Scotland. The recruiters from the various armed services used different methods to attract volunteers. Among them were representatives from the Polish Parachute Brigade, formed the previous autumn. They were dressed for action in jump smocks and armed to the teeth (indeed, one of them stood there with a commando knife clenched in his teeth), giving an enthusiastic spiel about how the paratroopers would be the first to return to Poland. They then demonstrated various phases of the jump from an airplane, simulating a landing with the proper roll and recovery, and then collapsing the billowing parachute canopy. This was followed by a lecture about the successes the German paratroopers had achieved, that we would improve upon these tactics, and apply them to the enemy.

This performance made an impression on many of the soldiers, but especially on us youngsters, who had survived the nightmarish experience of the "Soviet paradise." We could think about little except seeking vengeance on the Germans for starting the whole thing. It was the parachute that embraced my soul, along with those of many of the others there. My spirits were high. I no longer had any doubt - I told myself that I would become a paratrooper. I began a new chapter of my life, with no idea of the events that lay before me.

When I was still in Tehran, I wanted to serve in either a parachute unit or with the air force. Immediately after passing my medical and physical examinations, along with many others who dreamed of parachuting, I was sent first to Cowdenbeath, and then later Ealie on the Fife peninsula in eastern Scotland, where the training area for Polish paratroopers was located. And so began my long and laborious preparations, beginning with much physical exercise, before I could begin my parachute and combat training.

The Parachute Brigade was stationed in the county of Fife, with its headquarters in the town of Leven. The brigade's various sub-units were scattered among the various towns in the Fife's eastern peninsula between the Firth of Tay and the Firth of Forth. The symbol of our perseverance and efforts was crowned by our motto as paratroopers - we were taking THE SHORTEST ROAD back to Poland. This exactly described the objective of our commander, Colonel Stanislaw Sosabowski. This was our motto during our entire period of training. From the very first moment of the brigade's existence, it was to be a special formation under command of the Polish commander-in-chief for operations exclusively in Poland. Despite our desires to

Colonel Stanislaw Sosabowski, Commander of the 1st Polish Independent Parachute Brigade

see action in Poland, the Allied Supreme Command had other plans.

Initially the Polish armed forces in Britain had a surplus of officers in proportion to soldiers. When I arrived in Scotland, the 1st Polish Independent Parachute Brigade consisted of volunteers who came from different directions. The first were those who had managed to join the Polish forces in France in 1939 and early 1940. When France fell they found themselves in Britain. Among them were the sons of Polish immigrants who had settled in France or Belgium earlier in the century. A small number were volunteers from other Polish communities in the Western Hemisphere, including a pair of Americans who did not have a drop of Polish blood In their veins.

However, the core of the Parachute Brigade was formed by the so-called "Siberians." They, like myself, were those who left that Soviet "Paradise on Earth" with General Anders. In our company strength of 126 officers and men, we "Siberians" formed almost 80%.

A paratrooper had to be stronger, smarter, have greater stamina, and be much more efficient than the average soldier. As a result, our training was much more demanding. We were whipped into excellent physical condition as a result of countless hours of calisthenics, gymnastics, and daily long-distance marches and runs which prepared us for the next stage in our training at a place called Monkey Grove. The question, just what the devil was Monkey Grove, was bouncing around inside my head? Maybe we would actually be living among monkeys like Tarzan?

But then again, what could this possibly have to do with parachute training? In short order we found out what tortures would be inflicted upon our muscles and bones. Monkey Grove was a small corner of the estate surrounding the impressive manor known as Largo House, which was built in 1750 by some lord in the county of Fife. At first glance the estate looked like a large park. However, the physical training instructors had taken the majestic oak trees and installed various kinds of apparatus for perilous and daredevil exercises which were only for the audacious. There were trapezes, ropes suspended over water, ropes attached to the very tree tops that we would have to use to swing to another tree that had a narrow tightrope which led to another fiendish contrivance. And of course we had plenty of ropes to climb as if we were monkeys, which is how the place got its name. There were also various obstacles, such as narrow windows which we would have to dive through, deep ditches full of water that we would have to cross, and other objects that were guaranteed to be neck-breaking if one were neither fit nor careful.

Especially unpleasant was a barn where we would have to stand in the loft and jump into a hole in the floor. It was not a simple hole, but more like a barrel which had both ends removed.

Caricatures Of General Sosabowski And His Paratroopers
Training At Monkey Grove By Kazimierz Gramski

We would land on mattresses or bags of sand which were supposed to imitate what it would be like when we would be jumping from Whitley bombers - then the standard RAF aircraft for parachute jumping. It was easy to be injured parachuting from the Whitley. The paratroopers had to leave the airplane by jumping through a hole in the floor, which was lined with sheet metal. It was almost a meter between the floor where we were sitting, and the hole at the bottom of the fuselage where we emerged. This explained why we had to jump from the loft, through the barrel with no ends, and on to the mattresses. The biggest problem with this was that as your feet and legs came out of the hole, the aircraft's slipstream would catch you, and there was a very a good chance that either your head or your nose would get smashed into the wall of the exit.

But before we actually jumped out of an airplane, we had to learn and master all of the mysteries involved in the art of parachuting. For the moment, this was only theoretical, and was taught in a classroom.

After our muscles were sufficiently toughened, our reactions and reflexes sharpened at Monkey Grove, and our brains crammed with theory, our next step took us to a nearby parachute tower, designed by Polish engineers. It became a model for those used for parachute training throughout the British Empire.

Jumping from the parachute tower also had its place in our training. In many ways it was an imitation of jumping from an airplane. We had to climb circular metal stairs to a platform on the very top of the tower, which was similar to being on the roof of an eight-story building. The view was beautiful, but we were not there to enjoy the scenery. The wind whistled through the skeleton of the tower. My head felt scrambled, despite the fact that I tried to be cool, calm, and collected, and to appear steady to the man climbing the stairs behind me, my best friend — Tadeusz Boguniewicz. But deep inside of me I wondered why they chose me, and only me, to do this, when I did not have the foggiest idea of what I was getting myself into.

We were already wearing our parachute harnesses. The instructor at the top of the tower hooked an open parachute attached to a metal frame suspended overhead to my harness. He then told me to sit the edge of the platform with "the softest part of your body." I did not think with the harness that I had any place left to sit down. I was far forward with my legs dangling over. I was fairly pushed forward and just about ready to fall off when I heard the instructor bark, "Get closer to the edge!"

I made this minor adjustment and then like lightning, the word "GO!" struck my ears, and I went . . .

The sharp sounds of an instructor with a megaphone came up at me: "Curl up and shrink yourself, your knees under your chin, and keep your legs together!" There were those who thought themselves so brave that they would have jumped from a standing position, remaining at attention, but such craziness was not for me.

After several jumps from the tower we began to feel very confident in ourselves. This did not escape the attention of the instructor, who asked the question; "Who among you are going to be a statistic of those who have jumped to their deaths?"

The next step was to learn to deal with the parachute, not only in the air, but also during landing. When the paratrooper was in the air, the canopy was inflated with air, and difficult to manipulate, but on the ground, the sheer and silky fabric of the canopy seemed to have a life of its own. It was only too willing to drag us across the ground while resisting all our efforts to collapse it and free ourselves from its harness. It took little imagination to realize that one day we would have to do that while facing the enemy.

Ringway Airfield
January 18, 1943

After the toughening at Monkey Grove, and our jumps from the tower, we were considered to be ready for the next phase of our training; jumping from balloons, and then from airplanes. We were taken to the Parachute Training School at Ringway, an airfield near Manchester. There was a special Polish section at the school, where we had our own instructors. Before we could be considered to be qualified paratroopers, we had to make a total of seven jumps. The first two jumps were to be from balloons, and the last five from aircraft.

Those first two jumps from the balloon were most pleasant. It was very quiet sitting in the gondola, without the roar of aircraft engines, and we did not have to worry about the plane's slipstream, which we had been warned about during our classroom instruction. The gondola suspended below the balloon had room enough for an instructor and four jumpers. The winch cable then played out until the balloon was some 700 feet in the air. I was struck by how peaceful it was up there, surrounded by a blessed quiet. I was very calm and at peace with the world. It was a feeling that cannot be expressed with words, but has to be experienced.

I was jolted out of my introspection by the instructor's sharp voice:"Number two, action station. . . "GO!" I jumped through the hole in the gondola. Suddenly a resounding voice boomed through a loudspeaker on the ground: "Keep your legs together and prepare to land."

My second balloon jump that day was beautiful. After hitting the ground, I rolled and then collapsed the canopy. I gazed lovingly at the parachute spread sloppily on the ground before me. I felt buoyant and warm all over and just wanted to hug and kiss it as if it was the most beautiful woman on the face of the Earth. I had had my doubts about the parachute and did not have faith in it, but it did not fail me. I rolled it up, pressed it affectionately to my chest, and marched off to our assembly area. My joy knew no bounds, and I was sure that the rest of the world shared my happiness. Nevertheless, several times during the night I was shocked from my sleep when I heard that most disturbing phrase: "Number two, action station . . . GO!"

Now we were going to make our first jump from a Whitley, which did not resemble the balloon in any way, shape, or form. As we were driven to the airfield, I spent the entire ride concentrating on not thinking about what I was going to do. However, I was unable to free myself from those troubling thoughts. All that was going on in my mind was how I had arrived at this point, in a lorry being driven to the airfield from the exercises at Monkey Grove, the terror of having to launch myself through the "window," and all the bruises I had suffered. I gripped my parachute, but again I no longer had any faith that it would keep me from harm. I was now battling two emotions which had totally taken over — fear and foreboding.

Nobody was speaking, nor could we even look at one another. I realized that everybody was holding something back, keeping it bottled up. We all seemed to have become numb with apprehension. Somebody tried to relieve the tense atmosphere by telling a joke, but nobody was listening.

The ten of us climbed the ladder into the bomber. The fuselage of the Whitley was narrow and the ceiling low, so the jumpers had to get to their positions on their hands and knees with parachutes on their backs. The Whitley could only carry ten paratroopers, five on either side of the hole. I managed to take my place with my knees drawn up in front of me. The aircraft was still, filled with a complete silence --- that's the way it is among paratroopers before they jump. Our instructor, who had already completed dozens of jumps, told me that with every jump he always felt the same as if it was his first jump. I let the remark pass as if I had never heard it, not sure that he was telling the truth.

Just ahead of me was Edward Altheim, who was jumping as number three. He was always very talkative but now was silent, lost in thought. He quietly asked me, as a precaution, to check his parachute — and we all checked each other's parachutes and harnesses. The instructor was near the exit hole. His final words before the engines started were: "Do not lean forward before you jump, because you might lose your head. When you land, make sure you land with both legs, because otherwise you might have a serious misunderstanding with the ground."

With a nasty cough, one of the engines bellowed, followed by the other. There was no air in the fuselage, and now it became worse as the stench of exhaust was added to the smells of petrol, oil, and the sour scent of perspiration left behind by the others who sweated out their jumps, just as we were doing now. The cabin of the Whitley was an awful place. It turned out that the silence was a blessing when the aircraft's engines roared into life, with only the fabric skin of the fuselage between us and them. Most of the illumination in the dark cabin came from the hole below, giving the faces of those near the exit a ghostly look.

The converted bomber rolled forward as the engines roared. We left the ground. The sound of the engines was deafening. The seconds slid by, All of a sudden we heard the instructor shout, "Prepare for

Paratroopers Wait for the Jump Signal in the Dank Interior of a Whitley Bomber Sketched by Stanislaw Kowakczewski

79

action!" We all squeezed together toward the hole, and the red light went on. The instructor then shouted "Action stations!" and paratrooper No. 1 slid over and put his lower legs into the exit hole, and suddenly the green light flashed.
other side of the hole slid over, quickly put his legs into the hole and pushed off.

With the instructor's shout of "GO!" No. 1 immediately raised himself slightly on his hands and pushed off into the abyss. No. 2, on the other side of the hole slid over, quickly put his legs into the hole and pushed off. The instructor was off to the side, conscientiously watching everything going on, ready to step in should anybody panic or change his mind. But that seemed rather unlikely, as we were all squeezed together and pushing one another in the direction of the opening in the floor.

It was my turn, and I swung my legs over, pushed off, and closed my eyes. I felt the parachute harness tighten, and I opened my eyes to find that I was hanging in the air with a beautiful white canopy above me. This was a good sign: my reunification with the ground promised not to be fatal. I felt very happy and whistled cheerfully, as I am wont to do during such situations. I was suspended in space, and standing on air, while the ground curiously swung around below me. I was snapped out of my delightful mood by a loudspeaker booming from the ground: "Legs together, prepare for landing!"

My heels hit the ground, both at the same time, as we had been instructed. I rolled over, got to my feet, and as I started to collapse the canopy I looked in all directions, as if in combat.
started to collapse the canopy I looked around in all directions, as if in combat.

Over the following days I successfully completed my remaining jumps. The only problem was that on my next to last qualifying jump, I bounced a little too hard, and my nose suffered as a result. On January 23, 1943 I graduated and was now a Polish paratrooper.

We returned to Scotland happy and in a great humor as we had managed to accomplish what we had set out to do - - we had earned the badge depicting a diving eagle, ready to sink its beak and talons into the enemy. This we would now proudly wear above the upper left pocket on our uniforms, which would tell the world that we were Polish paratroopers.

Now qualified as paratroopers, we were to undergo further training. The majority would become parachute infantrymen, but others would be trained in various specialties. With my previous training in the Red Army, I naturally applied to become part of the brigade's newly established parachute signals company, stationed in the wonderful Scottish village of Markinch.

Our signals company commander, Captain Jozef Burzawa, was a cavalry officer. Despite the fact that he knew little about communications, we considered him a good officer who treated us as specialists.

But we were most fortunate that our radio platoon commander was Lieutenant Jan Wilk. He was a reserve officer who served in an armored battalion in 1939, before successfully reaching France. He was very intelligent, and knew much about radio communications.

Before his transfer to the parachute brigade, Wilk had been part of the Polish signals technical group in London, involved with the development of the suitcase radio. This device was a powerful radio transmitter-receiver, which could fit in a small suitcase. Designed for agents who were to parachute or infiltrate occupied Poland, it was adopted by British, American, and other allied special operations forces gathering intelligence or organizing resistance in continental Europe.

Lieutenant Wilk was short in stature, but solidly built. Despite the fact that he was very strict during training, he was very well liked except by those who sought to avoid responsibility, for whom he had no tolerance. Otherwise he was a kind and caring officer devoted to his work, his men, and Poland.

It was the beginning of 1943, and there was heavy fighting on many fronts. By spring, the Germans were beginning to lose the ground under their jackboots. They had been kicked out of North Africa, but before that, their surrender at Stalingrad was the pivotal battle of the war. From British soil, Allied bombers were over Germany night and day. We were under no illusion that the war was coming to an end, but we all instinctively knew that the Reich would eventually be defeated.

But in the meantime, we Polish paratroopers continued to train, and train hard. Our courses were long and complicated; both the theoretical and technical lectures required our full attention. Much was familiar to me, but the British equipment was much more sophisticated than what I had used in the Soviet Union two years before. But everybody in the Parachute Brigade was striving for one goal — being the first troops to liberate Poland. None of us imagined that the objective we worked and trained for so rigorously would not be accomplished; that there would be another outcome.

After morning muster and roll, apart from the specialized training unique to an airborne formation, we spent countless hours learning our jobs as signalers. Our days were taken up with becoming familiar with our radio and telephone equipment and above all, perfecting our skills in transmitting and receiving Morse Code. At the same time we had to become intimately familiar with the secret ciphers and cryptography we used in our communications, so that transmitting them became second nature.

Code. At the same time we had to become intimately familiar with the secret ciphers and cryptography we used in our communications, so that transmitting them became second nature.

A new group of men came to the company shortly afterwards. Among them was Private Tadeusz Boguniewicz. We had met earlier in Ealie, but in Markinch we became really good friends. He was very quiet and introspective, but we had very much in common. He came from Brody, a city between Lwow and Dubno. His father was a policeman in Volhynia who had been taken prisoner by the Soviets in 1939. Boguniewicz, along with his mother and sister, had been deported to the Gulag. He managed to get to Persia with them. When Tadek as sent to Britain to join the Parachute Brigade, his mother and sister were l iving in a refugee camp in Uganda. The quiet and thoughtful soldier spent writing letters to them which he sent along with most of his pay.

Tadeusz Boguniewicz

We became the closest of friends, even to the point of sharing the same bunk bed, with Tadek in the lower bunk, and me above. The company numbered some one hundred soldiers, billeted in a church hall. One could well imagine what this was like with all these people having to lived crammed in a tiny church hall, but remembering our "accommodations" in the Soviet Union we had no reason to complain. time was spent sending letters to them, along with most of his pay.

We became the closest of friends, even to the point of sharing the same bunk bed, with Tadek in the lower bunk, and me above. The company numbered some one hundred soldiers, billeted in a church hall. One could well imagine what this was like with all these people having to lived crammed in a tiny church hall, but remembering our "accommodations" in the Soviet Union we had no reason to complain..

Of course for everybody to live in such tight quarters, there had to be good order and discipline among the men. In our ranks we had some Jews, along with Ukrainians and Belorussians. They were all our comrades. There was a true brotherhood among us all, and from that standpoint it was indeed very pleasant being a member of the Parachute Brigade.

One thing did not change in Britain. In the morning, before our daily tasks and routines, we would have breakfast. Someone was sent to a local bakery to get bread. If we were lucky, they would have rolls, but they usually only had white bread. The food in Scotland was fairly poor and rationed, and we realized that the entire British nation was suffering shortages of all types. This was made worse by the German submarines, which tried their best to starve the isles into submission. Given the situation, and observing the stoicism of the civilians, we could not complain. However, the one thing we could never get used to was the bread. We were used to good rye bread, usually baked at home, which was whole grain and nutritious. In contrast, the bread that was served to us in Britain had no flavor to speak of. We felt as if we were eating cotton batting. It did nothing to fill our seldom satiated stomachs, but even for that, you needed a ration card.

With all the strenuous activities we had, we had to have calories to burn. For breakfast, we were served porridge, bread, margarine, jam, and of course, tea. The porridge was the traditional Scottish oatmeal, which was my favorite. Despite being rationed, jam was never in short supply. It was traditional, typical English jam, but how much of that could you eat, especially on bread that tasted like cotton? Among the many European nationalities who had escaped Hitler and made it to England were Czechs and Slovaks. Together, they formed a Czechoslovak Brigade. Their menu was obviously similar to ours, and those brigade's soldiers had a slogan, "We are not going to fight Germans for English jam."

Our technical training was demanding, but since we were paratroopers it was also imperative that we had to be in peak physical condition. Several times a week, we fell out in steel helmets and full combat gear, prepared for what we called "marching runs." The lovely Scottish countryside was a feast for the eye. The area was the birthplace of golf, and there were numerous courses in the

The Signals Company Passes in Review - Markinch

region. The grass in that damp, rainy climate was a beautiful emerald green, and sometimes, even now, I would see glimpses of these famous golf courses on television when championship matches were held there. But we were not there for sightseeing. We would start with a quick march for two miles, and then break into a run for the next two. This would go on for most of the day. By the time our hobnails were scraping the streets of Markinch, we were bone-tired, feeling as if we had been across the entire county of Fife, and we were as hungry as wolves.

Our dinner was mostly pea soup, and what tasteless bread was left over from breakfast. Rarely did we get enough to keep from being hungry, and our stomachs continued to growl. Still hungry, we would go to the high street where a long-time Italian resident had a fish and chips place. But again, it was all a matter of shortages. If the man managed to obtain fish and potatoes, then we had to be sure that we were not suffering from a shortage of shillings -- fish and chips were expensive on a common soldier's pay. Due to a shortage of newspapers, one sheet had to suffice to form the paper cone when the piping hot delicacy was served to us. It was delicious and filling, but the lower left sleeves of our battledress blouses were darker than the rest, because of the grease that came through the paper cone and rolled down our wrists.

Once a week, during our free time, Tadeusz Boguniewicz and I would spend an evening engaged in private English lessons. Our teacher, a certain Miss Ping, was a retired schoolteacher, and was, so to speak, an old maid. We had managed to tell her about what we had been through in the Soviet Union. She understood this well, having great compassion and sympathy for us.

The matter was becoming rather urgent, as our spoken English was very poor. The only way that we new arrivals could try to arrange a date with a Scottish girl was with wild gestures and the broken English litany; "You go, me go, bus go, Glasgow." These were precious evenings with English lessons, interspersed with a cup of

I POLISH PARA BRIGADE SIGNAL COMP, SCOTLAND-MARKINCH-1942

tea and piano recitals by Miss Ping, who adored Chopin. This gave us a bit of variety in our rather monotone lives as soldiers. In six months, my English was good enough to be able to ask a girl to dance without having to resort to gesticulation and pantomime.

Then in April, 1943, we received shocking news. German radio announced the discovery of mass graves near the village of Katyn, Russia. The Germans claimed that they had found the corpses of 10,000 Polish officers. All had their hands tied behind their backs, and had been executed by a single bullet to the back of the head. The Soviet government, of course, denied responsibility, though we all knew well what they were capable of. Many Polish officers and civil servants had not reported after the amnesty. My comrade Boguniewicz was numb. He knew that he would never see his father again.

In May, 1943, the parachute signals company moved to its summer camp in Tentsmuir, near the RAF Leuchars airfield. The facility was well chosen. Located where the Firth of Tay went into the North Sea, it was far from populated areas and had few roads. The terrain was ideal for battlefield training. Polish paratroopers would leave with unforgettable memories, some of them very humorous, and others, unfortunately, very sad.

As Tentsmuir was fairly far north in Scotland, and the summer days were long. We were quartered in Nissen huts, made from corrugated sheet metal in the form of a half cylinder, with the ends sealed with brick or lumber. They were inexpensive to produce, very versatile, and could be erected quickly. The Polish soldiers called them "barrels of fun" as an ironic acknowledgment of the Beer Barrel Polka, which was so popular then. These simple tin structures became functional homes, workshops, offices and recreation spots for thousands of soldiers and airmen in WWII. Nissen huts had concrete floors and usually had a small, inefficient coke stove for warmth.

However, these temporary structures had no latrine facilities, which were a long walk in the chilly night. And similar to the battlefield, washing facilities proved to be non-existent, which made living conditions for the men crowded into the Nissen huts even more trying.

We had calisthenics every morning. Everybody was required to participate. This situation gave birth to hundreds of ideas and schemes in order to get clean. As soon as our morning exercises were over, a great race began toward the drainage ditches with the objective of washing up. All of this reminded me of the daily competition to find a good spot to wash in the Oka River while I was in the Red Army, with the difference that the water was much colder. After a breakfast of porridge cooked in a field kitchen, much of the day was dedicated to technical training, despite the fact that everybody's favorite time was at the rifle ranges.

But field training was not always sweat and toil. Lance

Sergeant Edward Sobieralski, the leader of our WS 22 team, was a talented accordion player. This veteran of the French campaign was able to get the most wonderful sounds from his button accordion. His beautiful melodies flowing through our bivouac comforted us, and the pressures of training and the fatigue from our long and hard hours of training for battle began to evaporate.

Czeslaw Gajewnik With A Young Scottish Friend

Another close friend was Czeslaw Gajewnik, a fellow "Siberian." He grew up in a small village in the Lublin region, and always had a smile on his face. It was impossible not to like him, and he was devoted to his work and card games.

We began having closer cooperation with the British paratroopers. Our comrades-in-arms from the British 1st Airborne Division presented us with a banner. The presentation ceremony was attended by all the Polish paratroopers who were at Tentsmuir. But it turned out that of all of the events during our training at Tentsmuir there was one that would sadly overshadow all other all other memories of the place.

On June 14, 1943, a group of British paratroopers from the 8th Parachute Battalion came from their base at Bulford Camp in southern England to take part in a combined training exercise with us. With many VIPs on hand to observe, the British paratroopers had flown from southern England in ten Whitleys to conduct a parachute assault against us. Through some kind of error, a stick of British paratroopers jumped from their Whitley, and into the Firth of Tay, where all but one drowned. The British paratroopers

Joint Maneuvers With British Paratroopers at Tentsmuir

remained with us for a week, conducting tactical exercises. However, as a result of the tragedy, the training was half-hearted under the pall of the recent deaths.

After many months of constant preparations, our brigade, from the standpoint of individual training, was in outstanding condition. But there was another matter that few in our brigade were aware of: in 1943 the commander-in-chief of British airborne forces, General Browning, approached Colonel Sosabowski with a proposition that he take command of a joint Polish/British airborne division. Knowing that the Polish Parachute Brigade was designated to participate in the liberation of Poland, Sosabowski declined the British offer. General Browning was deeply offended by this refusal. Perhaps this incident was the cause of the deplorable situation between them that they would find themselves in later.

Then, on July 4, 1943, Captain Burzawa called us all together before breakfast. He made the tragic announcement that General Władysław Sikorski had been killed in an airplane crash in Gibraltar. This was the worst imaginable news that could be given to us. Just as BBC radio broadcasts provided a beacon for the hopes of the people of occupied Europe, to the Polish soldiers fighting side-by-side in the ranks of the Allied armies, the person of General Sikorski personified their hopes of returning to a free and independent Poland. Although he had his detractors in some sectors of the Polish government and military circles, it was through his leadership and efforts that we "Siberians" were freed from Stalin's yoke. Had it not been for him, we certainly would have joined the thousands of young Poles from the Borderlands who perished from exhaustion or starvation in prisons or Soviet labor camps. This was a shattering blow, coming on top of many others. We had great faith in him, knowing instinctively that he had the talent to be Poland's voice among the Allies, including the one who had been our enemy, and which now had great influence with Churchill and Roosevelt.

1944 also saw us bidding farewell to the Whitley. Instead of jumping out of a hole in the belly of the "flying coffin," as we called it, we were introduced to the American C-47 aircraft, which the British called, the Dakota. This would now be the aircraft that would take us into combat, so we would all have to be trained in jumping from the Dakota. We went in groups to Bulford Camp in southern England for our conversion training. I must admit that there was quite a contrast between the two. We sat in seats on the Dakota, instead of on the floor, and had windows to look out off rather than the Whitley's gloom. When it was time to leave, you left through the open door of the airplane like a gentlemen.

Stamford/Peterborough, England
July 1944

Our brigade was put under the command of the British 1st Airborne Division and transferred to the Stamford-Peterborough region of England. The brigade moved by rail, except for advance parties which had prepared facilities for their units. The signals company was assigned quarters near the quiet English village of Easton on the Hill. We occupied Nissen huts recently vacated by American soldiers. But we, like the other Polish paratroopers, did not have the luxury of sharing the village's quiet life. All units were busy and the signals company was no exception.

The signals company provided the 1st Polish Independent Parachute Brigade with three types of communication; telephone, radio, and messenger. The last, depending on the situation, operated on foot, with bicycle, or motorcycle. The telephone platoon operated switchboards at brigade headquarters and laid and maintained telephone lines to all of the brigade's rifle battalions and sub-units. The tools of the trade for the radio section were:

The WS (wireless set) 38 - a

WS 38

small man pack set for communications on the platoon and company level. It was a voice-only transmitter-receiver, and usually used a throat microphone.

The WS 18 was the standard wartime British Army man-pack radio, and was designed for short range communication working in forward areas on the company and battalion level. It could be operated by one man in a fixed position, but while moving, it was usually carried on the back of one soldier and operated by another. It was power by dry-cell batteries. We were re-equipped with the WS 68P before going into action. This improved set did not differ externally from the WS 18, but operated on a lower frequency for use by the airborne forces.

Polish Paratroopers In Holland With WS 68P

The WS 22 set Showing its Components. I have the Morse Key on my Knee

The WS 22 was for communications at the battalion and brigade level. This radio was capable of both voice and Morse transmissions. It came in two components, the radio itself, and a power supply. It was heavy, and used wet-cell batteries which caused the it to usually mounted in a vehicle. As paratroopers, we did not have that luxury, and we were also trained to operate it mounted on a wheeled trolley. The batteries were recharged in the field by a petrol-driven charging generator.

Lastly, there was the WS 76. It was a shortwave Morse transmitter/receiver. It had tremendous range and provided a direct link with the Polish headquarters in London.

The WS 22 On Its Trolley
From Left to Right : Lance Corporals Stanislaw Muras, Edward Altheim, Stefan Korzepa And Waldyslaw Kuczewski, Who Was Later Killed In Action.

In addition to providing communications for the brigade, the signals company supplied and maintained the signals equipment for all of the brigade's sub-units, along with training the operators. This made for a very busy schedule and a large work load for all of us. The signals company was truly blessed as the people who served were almost without exception intelligent and hard-working team members.

The WS 22 As Operated On Its Trolley

    We did a few jumps when we were in England but that was mostly conversion training to learn how to jump from the Dakotas, rather than from the horrible Whitley bomber. We did not do much jumping in Scotland; it was both a matter of the British unable to spare aviation fuel to allow us to practice, and General Sosabowski also did not want to unnecessarily lose any men to the risks of parachute jumping.

    As we were approaching the final phases of preparing for action, we took part in several simulated combat jumps with full equipment. During one of these there was a tragic accident. Before the parachute jump, two of the Dakotas collided in midair. Twenty-six Polish paratroopers were killed, along with the American aviators. Among the dead were three of my friends. We were witnesses to this violent accident, and it seemed as if a dark and depressing cloud descended over our brigade.

    During our next training exercise the entire brigade would be making a night jump. I came down rather hard, cracking my lower shin bone. I reported my injury to Lieutenant Jan Wilk, the commander of the company's radio platoon. He asked if I was able to walk, and then if I could still carry out my duties. I told the lieutenant that I was feeling pain and had some trouble carrying loads. However, I told him that I would be able to manage. This was part and parcel of the paratrooper spirit, that we would accomplish our mission to the best of our abilities and to conquer any adversity. Though I did not know what would happen to me afterwards. I admired and trusted both my comrades and my officers and feared reassignment. The men in my platoon were like brothers to me, especially Tadek Boguniewicz.

On the night of August 1, 1944, the Polish broadcast on the BBC told us of the outbreak of the long-awaited Warsaw Uprising. This news was initially received with joy, as it appeared that the end of the war might be near. But as the days went by, there was no change from our busy preparations for combat operations, and no hint that we would be taking part in the promised liberation of Warsaw. Reports in the Polish press indicated little progress by the Polish underground army and the apparent indifference of our British and American allies.

Had we not been working around the clock getting ready to go into action, we would certainly have had nervous breakdowns. Our company's lorries were leaving for the continent where they would be ready, along with the brigade's other vehicles, to meet and reinforce the parachute and glider elements. They were packed to the limit with heavy equipment that could not be dropped by parachute or taken by glider, along with spare parts and replacements for items that would be lost in combat.

On August 13, 1944, our brigade received definite orders for a combat jump. Our objective, along with the British 1st Airborne Division, was to seize and hold Rambouillet, outside Paris. But nothing came of this operation as the swift advance of the Allied armies overran the objective and the mission was called off mere hours before we were to take off.

A short time later the brigade was again prepared, and buttoned up to the last button for another mission, only to have the jump rescheduled, and then finally canceled. Waiting to take off for these numerous missions, only to have them canceled, was nerve-wracking and discouraging. Not only was there the tension of facing the great unknown of combat, but the heated preparations the exhausting labor of packing and then unpacking our equipment from parachute drop containers or gliders began to wear us down.

Across the continent in Europe, the uprising in Warsaw did not turn into the swift victory that we had hoped for. It was now stalemated and had been going on for over a month. Despite our near exhaustion from many hours of work, we tried to get as much information as we could. However, there was very little. We wondered why we were not jumping into Warsaw. After all, this was the mission our brigade was created for, and it seemed that none of our allies would or could do anything to help. All we knew was that Warsaw was burning, and the Red Army did nothing but watch from across the Vistula River.

During the first week of September, we were restricted to our camp, again surrounded by armed guards to keep us isolated. We were briefed that our brigade, along with our comrades from the British 1st Airborne Division, were to take and hold bridges over the Maas and Waal Rivers in Holland. This was called Operation COMET. Again we worked like bees to get ready. Because of the weather, we did not take off as planned on Sunday, September 10. The following day, the weather was good, but the operation was canceled. We were told not to relax, that we had received orders for another operation, named MARKET GARDEN.

Operation COMET was expanded to have three times the number of paratroopers to take part in the invasion of Holland. Our brigade was to be attached to the 1st British Airborne Division, but now our objective was the bridge at Arnhem. This was over 100 kilometers behind enemy lines. The plan was for two American airborne divisions to land and seize a corridor for the British land forces to relieve the 1st Airborne Division in four days.

All of this seemed simple on paper, but with so many gliders and paratroopers to be delivered, there was a shortage of aircraft. Our brigade would go in on the third day of the operation. Our heavy equipment was to go in by glider on the north side of the Lower Rhine River, while the majority of us would parachute south of the river. We were then to cross the bridge, which was to be taken by the British on the first day of the operation, and occupy defensive positions.

The days flew by, and we tested and retested our equipment, cleaned our weapons, and made certain that our individual kit was in battle-ready condition. But we, as all Polish people around the world, remained distracted by the events in far-away Warsaw. Our countrymen there had been fighting bitterly for over six weeks, while the Germans destroyed the city block by block. As soldiers we knew that we had to carry out our orders, but the fact that our brigade was formed to be the first to return to Poland, and here we were, going to liberate others while our compatriots were locked in a death struggle with one enemy, while the other just watched and waited, was agonizing.

We went through the same motions that we had several times before. While we signalers again prepared ourselves for battle, and were issued dehydrated emergency rations, wound dressings, and morphine. Then came live ammunition, mines, and grenades. We were told to take as much ammunition as we could carry. Then we were issued parachutes and were armored vests.

Part 5

Easton on the Hill, England

On September 16, we received strict orders that we were not to leave our billets. Sunday, September 17, 1944, dawned with absolutely beautiful weather. It was one of those gorgeous September days that made one glad to be alive, despite the fact that there was a war. There were no religious services that day. The brigade's chaplains had previously visited the units and gave the men absolution. Later that morning, the skies were filled with Dakotas. We now knew that Operation MARKET GARDEN was on, and it would be our turn in two days to take off and join the fight in Holland.

Taken Before Leaving Scotland, 1944 Visible Above The Are The Badges Awarded When I Had Qualified As A Signaler, And As A Paratrooper

One by one, we all took stock of our situations. We knew we were going into action, and now there was a minimum of happy banter as everybody made their final preparations. Tadek Boguniewicz and I had exchanged all our family information should one of us have to perish. We were not alone. I believe that soldiers had been doing the same long before the birth of Christ. My leg would throb from time to time, telling me that I still had a broken bone. But my leg concerned me less than what we would expect after landing. Would our team be able to provide adequate communications for our commanders while under enemy fire? Would I ever see Jozefa again? Those thoughts were persistent, and evr present.

Early on the morning of September 19, British lorries came and picked up us and our equipment. Left behind was 2nd Lieutenant Leonard Buren, the company's supply officer. He was in charge of liquidating the camp. The villagers waved goodbye to us as we made our way to Saltby Airfield.

We had our on jump smocks, and strips of burlap woven into the nets on our steel helmets. We were loaded down with ammunition, picks, shovels, body armor, radio equipment, spare parts, batteries, radios, weapons, explosives, mines, grenades and food for three days. Our personal equipment, with parachutes, weighed some 140 pounds, and prevented any free movement. And over our paratrooper jump smocks we wore inflatable life vests called "Mae Wests" by the British to save us from drowning in the English Channel in case the plane went down.

It was very foggy at the airfield where the Dakotas, set in rows and numbered, were ready to take off. We looked for the plane with the number "102" chalked next to the door. While we attached the equipment containers under our airplane, the stickmaster checked the cables, the external container racks, and the jump lights on their aircraft and confirm the stick rosters. The soldiers then slung the drop containers with radio equipment, arms, and light motorcycles under their airplanes. Folding bicycles and trolleys were loaded into cabins near the door where they were thrown out just before the stick jumped.

After we finished all the preparations, we tried to relax near the airplane while some of the officers in our stick spoke with the American fliers. The hours crept by until the afternoon, when we were told that the drop had been cancelled. We were not very happy, which was a contrast to the American flyers who were relieved that they would not have to face the enemy for another day.

Polish Paratroopers at Saltby Airfield

When the British lorries dropped us off at our old billets, we were greeted by a bewildered Lieutenant Buren. The billets were empty and everything was stacked outside. Blankets and mattresses were pulled from the piles, and we were fed cold sandwiches as the kitchen was packed up.

The following morning we were again driven out to Saltby. The fog was no different than during the previous day, but the atmosphere was completely different, and there was a palpable tension in the air. General Sosabowski had all of the officers running around and reporting to him. We later found out that our orders had been changed, along with our drop zone.

The new maps issued to the officers told us that we would be dropped to the east of a village named Driel, and that our officers would give us further instructions when we were on the ground. Now, instead of landing outside of Elden, just south of the Arnhem bridge, we would now be jumping somewhere else. We were very quickly briefed about the characteristics of our new drop zone, the new assembly areas for all of our units, and the changes in our combat orders. We returned to the aircraft and continued to wait. This news caused much speculation, and little of it happy. Everybody was reading between the lines of our new instructions. There had been minimal news from Holland, and now there was definitely something wrong. All the newspapers could tell us was that the ground forces' advance was going well, but oddly there was little mention of the British paratroopers at Arnhem, and we now knew that something bad had happened during the battle for the Arnhem bridge, but what? What could we expect when we got there?

As takeoff was postponed for an hour, and then another, in hopes that the fog would lift, the nervous atmosphere could be cut with a knife. Again we were told to go back, and that we would have to try again tomorrow. Back at Easton-on-the-Hill, Lieutenant Buren was incredulous when he saw us arrive. Again we had to help ourselves to whatever we needed for another night from the equipment that the supply officer had packed back up during the day.

The soldiers were given the opportunity to sleep, but few of us could. The brigade's officers did not have that option; the delay had them huddled and refining the new operations plans that had been hastily cobbled together when the change in drop zone was received.

When we arrived at Saltby on the morning of Thursday, September 21, the airfield remained wrapped in heavy fog. Around 2 in the afternoon, the fog still had not lifted, but meteorological reports guaranteed that there would be sunshine over Belgium and Holland.

But there was something else bothering General Sosabowski, and that was the lack of any concrete information about the situation that the 1st British Airborne Division and the Polish paratroopers who went in by glider now found themselves in.

We boarded the aircraft, which were lined up at the edge of the runway, one after the other. The minutes passed, and began to add up to almost an hour. The sky and the airfield remained enveloped in thick fog. In the distance, we heard aircraft engines bursting to life. Will we finally take off? When the engines on the Dakotas fired up, we know the die has been cast.

We taxied, and then the engines roar into a crescendo, and we lift off, and into the thick fog. We were all silent, through the windows all we could see was a dark gray veil. Finally, we broke into sunshine, but when we looked out the windows, we cannot see the ground, just a fluffy carpet of white clouds. I saw planes all around ours, flying in tight formation. I began thinking that we would have to loosen our formation due to the antiaircraft fire which would undoubtably greet us when we crossed the Dutch coast.

Each time we returned from the airfield, we were in a state of unbelievable tension. It was a combination of wondering what was happening when we had not received any news or information from the front, and anticipation and fear, which became worse with every minute we waited to take off. It was not surprising that the nights were a real nightmare, despite the fact that we could not sleep because of the day's events.

I looked at the faces of my comrades, and they were all very serious. Someone tried to break the silence, and said something funny. However, the joke brought no laughter and the paratroopers remained silent and locked in their own thoughts.

I paid no attention to the time, but I somehow felt that we should have been over Europe by now. Looking down through gaps in the clouds, and saw that I was correct. I could see fields, buildings, and scenes of destruction left by the war. Only the pilot knew where we are. Suddenly small black puffs of smoke, along with violent thumps shook the airplane. Without a doubt, these explosions were from anti-aircraft guns welcoming us to enemy territory.

We're seated on benches that run along the sides of the fuselage, but we have to brace ourselves try to stay in place as the Dakota bounces and jolts. I am resigned to what will be will be on the battlefield, but begin praying silently to myself that our plane would not be shot out of the sky. I ran my hands along my parachute harness, but it did not give me much confidence if our Dakota was shot down or caught fire.

While speaking about the possibility of being shot down, our brigade had taken measures in case of that eventuality. Members of the command group and other specialist units, such as our signals company, were divided among different aircraft. This was done in case an aircraft was shot down that it would minimize casualties among the commanders or specialized troops, such as we signalers who had undergone long and valuable training.

We could feel the airplane beginning to slow down and knew that we would be over the drop zone shortly. Then came the long awaited moment when the red light went on, meaning "ACTION STATIONS." All of the paratroopers stood up and checked each other's equipment. In a moment the green light flashed; GO!

From that moment everybody reacted automatically. Our stickmaster released the containers slung under the aircraft, one of which carried our radio equipment and batteries, and then he went out the door. Then, one after another we followed him. In the first moment, I was slammed by the wind, and in an instant began to fall like a rock until I felt a sharp tug on my harness. My parachute opened beautifully and I was suspended under its canopy.

I released my kitbag as machine gun bullets and antiaircraft fire ripped through the sky. With all of the lead and steel that was flying around, I was afraid of being shot out of the sky like a duck. The seconds passed as if in slow motion as I put my feet together for landing and tugged on the front risers of the parachute. As we were taught, this spilled some air out the back of the parachute to speed my descent to the ground before the German machine guns got me.

There was no wind, it was an excellent jump. I suffered no injury as my boots landed on the soft soil of a field of leafy vegetables. However, I saw that I was fairly lucky and some of my comrades had difficult landings. Some of the paratroopers landed in water filled drainage ditches that crisscrossed and separated farms and pastures. All around me paratroopers were landing, collapsing their parachutes, unbuckling the harnesses, and moving with great speed and energy. Time was short, and I collapsed my parachute, and twisted and then hit the quick release buckle of my parachute harness. I shed the sleeveless over smock that kept my equipment from fouling the parachute lines, and then glanced upwards. My eyes took in all that is going on above me. I saw of the Dakotas, now empty of paratroopers, get hit with tracer bullets. It caught fire, and left a trail of heavy black smoke as it lost altitude until its final and violent meeting with the ground.

I looked around for my kit bag, which I had lowered on the end of a line while I was still in the air. I was so distracted by all that was going on around me that I had not noticed that it had landed right beside of me. Most important was the fact it had not landed in a water filled ditch. I opened my kit bag and the first thing I did was assemble my Sten gun and insert a magazine. Now I had something to defend myself with.

I looked around to find the others from my radio team, and the orange parachute that marked the container that held our radio set. Boguniewicz was the first to reach it and was unsnapping the brackets. Close by was our two wheeled trolley, identified by its orange parachute. The radio set was very heavy, and the trolley had relatively small wheels. Two of us were pushing the trolley, and two others pulling it by the ropes attached to the front. It was very rough going over the muddy fields which were bisected by many ditches. It did not help that the broken bone in my foot was starting to hurt, but I ignored it and carried on.

The machine gun fire had stopped, but mortar shells were still landing on the fields now covered with discarded parachutes and odd bits of equipment. There was movement everywhere. We saw some of our paratroopers being carried off of the fields, on their comrade's backs, cradled on rifles, or even wheeled on trolleys to a dressing station set up at the edge of a road.

Our assembly area was the same as that of the brigade

headquarters which we were assigned to. We were to meet at the Baarskamp Farm. Finally reaching a road, we were able to move the radio without a struggle when we saw a group of officers. Among them was General Sosabowski, who was speaking to a Dutch woman. She was amazed that we were Polish, and not British paratroopers, which the villagers knew were fighting on the other side of the river. She spoke English and reported where the Germans were, and provided other important information, including the fact that the ferry our brigade was to use to cross the river was gone. She was a great help, and General Sosabowski was very happy with this information. I later learned that this was Miss Cora Baltussen. She and her family would forge enduring bonds between we Polish paratroopers and the village of Driel, which remain to this very day.

While all of this was going on, I immediately got the radio working and began contacting the other units. As a unit reported, our runner would take that information to the general. But we had no contact with parts of the 3rd Battalion or the entire 1st Battalion. Also missing were our signalers who were with those units, and our company commander Captain Jozef Burzawa and his group. General Sosabowski was no longer in a good mood, and told us to keep trying to make contact with the missing units. Our Morse signals went out into the ether without a reply. Eventually we were told to stop and save the batteries.

While paratroopers went off to find boats, the headquarters WS 22 set operated by Corporal Ignacy Sas kept trying to make contact with our signalers with the British 1st Airborne Division, but without success. Orders came down that we were all to prepare to go to the dike and wait for an opportunity to get across the river. Twilight was falling as we moved out as quickly fell. In addition to our trolleys, we now had horses and wagons captured from the Germans to carry the wounded, and heavy equipment.

The night was cold and punctuated by the sounds of machine guns and explosions coming from across the river. We were now doing one of the characteristic things that soldiers did, we waited. There was little conversation, with most of the paratroopers wrapped up in their own thoughts and fears. Every so often Corporal Sas again tried to contact the men across the river. Lieutenant Tadeusz Rakowski, the company's deputy commander, was now in command. Everybody, and nobody as much as General Sosabowski, wondered what had happened to all of the missing men. Knowing the hot reception we had received in the air made us shudder at the thought that many of our comrades were dead. It was better just to stop thinking and trust that God would get us through.

As midnight approached, Lieutenant Wilk came to us with the news that Captain Zwolanski had swum across the river. The Polish liaison officer, who had gone in with the 1st Airborne Division by glider, reported that the British were going to provide boats and rafts for the brigade to get across. We were told our 3rd Battalion was to cross first, and our company would follow along with the brigade's headquarters. From time to time artillery fire from across the river went whistling over our heads, exploding in the distance.

It was becoming light as orders come down that we were to return to Driel. General Sosabowski established his headquarters in a solid brick farmhouse. The infantry moved out through the orchards to take defensive positions. I was exhausted from lack of sleep and the tensions of the past few days. My stomach was growling but there was no thought of either food or rest, as we had to dig in as quickly as possible. General Sosabowski ordered that we were to prepare a hedgehog defense of the Dutch village.

All around, men were bent over digging slit trenches, except for the telephone platoon which was stringing wire to all of the positions. We have no idea when more, if any supplies will arrive, and have to be very frugal with our batteries. We had spent another night in a row we got little or no sleep, with the last being real nightmare. General Sosabowski personally inspected our defensive positions, and when he moved on we managed to gnaw on hard army biscuits, and grabbed a few apples from the trees, but when would I be able to grab some sleep?

Between 8:30 and 9 o'clock a lot of things happened. Corporal Sas finally made contact with the Polish radio team in the 1st British Airborne Division's headquarters across the river. As men finished digging their foxholes, they were sent out to the drop zone

British Armored Cars Arrive In Driel
The Bare-Headed Soldier Is Lance Sergeant Konstanty Zmaczynski
From The Signals Company

to gather up any containers or equipment that had been left there. Then, to our surprise, a British armored car and a scout car drove up to our headquarters. Some of the our signalers went out to talk to the British soldiers who turned out to be from a reconnaissance unit from the British XXX Corps.

This happy affair came to an end when small arms and machine gun fire was heard coming from the drop zone. We got back to our foxholes and readied our Sten guns and grenades, but only as a last resort. As signalers, our mission was to maintain communications and protect our equipment from any harm, we would turn to our weapons only if there was no other option.

The sound of fighting increased from the south, and reports came in that the village was under attack by the Germans, who were supported by halftracks. Artillery began whistling into Driel. The church tower in the center the village proved itself to an excellent aiming point. It did not take long before the headquarters building was bracketed by explosions. Tadek Boguniewicz and I pressed ourselves with all our might into the floors of our foxholes.

During a pause in the shelling, Lieutenant Wilk, fearing that the enemy was using radio location equipment, ordered that we move the radio set to another location. We moved the radio some 30 meters away from us, and using a remote cable hookup, kept the radio in service. Boguniewicz and I were sitting on the edges of our foxholes when we saw Lieutenants Rakowski and Wilk coming to check on us. At that moment there was a flash of orange light in front of me, and instinctively I dived into my foxhole and buried my face in it. I then suddenly felt a strange vacuum in the air, and then pressure, as if a large hand was pressing me into the ground. I heard an explosion, and then another and Boguniewicz utter, "Oh God, is this the end?"

I was covered with earth and took stock of myself and realized that I was alive and not wounded. I looked over to see my friend Tadek had disappeared into a pile of freshly churned earth. I saw the bottom of his boots under the heap of dirt. I frantically began digging with my hands, which found his, and I pull him out. Half deafened by the explosion, my closest friend told me that he thought that this was the end. Neither of us were wounded, but we were certainly shaken. I saw medics putting Lieutenant Rakowski onto a stretcher. Another man was wounded, along with Lieutenant Wilk, who despite suffering injuries to his face and eye, took over command of the company.

The fighting at the perimeter of our hedgehog went on sporadically through the afternoon, as did the German bombardment. At one crucial point General Sosabowski mounted a lady's bicycle and led a British armored car into to repel the attack. By late afternoon we no longer heard the sounds of fighting to the south. A long artillery barrage followed, which must have been to cover the withdrawal of the attacking enemy forces.

Driel
September 23, 1944

When there were airdrops by British transport aircraft we were treated to a beak in the artillery bombardment. The German gunners, apparently afraid of Allied fighter planes escorting the transports ceased fire. But the enemy anti-aircraft guns went to work and filled the sky with tracer rounds and the bursting shells that had greeted our arrival in Holland. It was impossible not to watch the slow but steady transports hold their course through the hellish fireworks display as parachutes of many colors, each designating the content of the container or pannier, suspended below. The entire scene would have been a spectacular display if the majority were not falling into enemy held territory.

Even more tragic was when the German gunners hit their target. A four-engine Stirling was trailing smoke as it lost altitude crossing the river. We knew that the plane would be in the air much longer, and had no doubt that it was too low for parachutes to be of any use to the aircrew. Another wave of aircraft appeared, just east of us. This time it was Dakotas, and the enemy anti-aircraft gunners again filled the sky with tracers and explosions. Again, multi colored parachutes blossomed, and one of the Dakotas wing caught fire. We watched helplessly as the fire spread and the wing broke off. The plane tumbled onto the meadows just north of the river.

At twilight we received reports that British tanks carrying infantry had arrived at Driel. It seemed that the entire village sighed with relief that the land forces had finally arrived, and that the tanks would strengthen our defenses. The men of the 3rd Battalion formed up, and marched off to where we had waited for boats to get across the river the night before. They were to go across the river that night. Everybody else was warned to be prepared to go across at a moments notice.

Reinforced by British tanks and infantry, we received orders that one man would stand watch for an hour, while two others were to get some sleep. As we were signalers, not riflemen, we did double duty having to have someone to man the radios at all times. During my turns at duty, there were flares lighting up the area behind the dike along with explosions and machine gun fire. Whatever was happening there, I was glad to be where I was, and even managed to get a few hours of sleep.

Before dawn the soldiers of the 3rd Battalion who did not get across the river returned to Driel and re-occupied their old positions. The crossing was accomplished with a few rubber dinghies, and happened under heavy mortar and machine gun fire. We learn that only some 50 paratroopers from the 8th Company managed to get across the river.

There was more of a British presence in Driel now. Other than the infantry and tankers, there were a few men wearing red berets.

One of them was a British signals officer with several men. He asked for inclusion of one radio set to our net, and was given permission, as ours was the only reliable link with the 1st British Airborne Division. He also brought us the first news of what happened to the missing men from the 1st and 3rd Battalions. Rather than falling victim to the enemy, they had returned to Britain because of the weather. If everything went right they were supposed to parachute into the area secured by the Americans to the south of us.

We were experiencing a shortage of food and ammunition. The brigade's order of the day emphasized that ammunition be conserved, and that the enemy be engaged only if we were certain that we would not miss. Most of the rations that we had carried had been eaten, and we availed ourselves of the apples growing all around us. Some of the men had managed to cut chunks of meat from the cows that had bee killed by the bombardment. The thought of the beef was most appetizing, but how to cook it? Any rising smoke would have drawn enemy fire. We were again heavily shelled during the day.

Driel
Night of September 23/24, 1944

It was dark when ammunition and rations from the 43rd Division were distributed, along with news that we were to pack up and prepare to try to get across the river again. An hour before midnight, we began a 2 kilometer march to the same place where we waited for boats the night we landed and the 8th Company had crossed the night before. The 3rd Battalion would be the first to cross, followed by the parachute lift of the anti-tank company, who would join their comrades and the guns that had arrived by glider. Next to cross would be the signals section of Brigade Headquarters, along with our signals company, followed by the rest of the headquarters.

Moving our trolley with the radio in the dark was not easy given the soft soil. When we arrived at a farmyard we were divided into boatloads of fourteen men. Allied artillery fire from the south passed overhead, and hopefully landed on the German positions to soften them up. This stopped at midnight. As if in spite, the Germans replied with mortar fire, which fortunately landed nowhere near us while we made our final preparations behind the dike.

In a half hour, we heard British lorries approaching our positions with their lights out. Despite their best efforts to remain quiet, we heard the noise of the assault boats being unloaded, and then the grunts and groans of soldiers carrying them closer to the waiting paratroopers. Officers then moved down the line to the groups of men divided into boatloads, telling them that the boats that had arrived could carry only eight men, not the fourteen which they were told that the boats would carry. We had to do this in the dark, in conditions where you could not see your hand in front of your face.

The boats were heavy, each weighing over 300 pounds. The soldiers had to carry these up the steep side of the dike, and cross the road which ran on top of it. Then they had to go down the other side of the dike, and cross a boggy pasture which ran 200 meters before reaching the scant cover of a low dike, less than a meter high before reaching the edge of the river. Though we were spared from having to carry the boats, we were responsible for getting our radio set, its equipment, and the trolley across the river.

It was 3 o'clock in the morning when the sappers and the infantrymen began moving the boats over the dike. As soon as they had reached the road on the top of the dike, flares suspended from parachutes appeared. They lit up the meadows leading to the river as if it was daylight. In seconds Spandaus started raking the area, followed by mortars and artillery.

We were to follow with the second waves of boats, some of which were being carried by some of our signalers. Moving the trolley and the precious WS 22 radio up the steep and muddy dike was a nightmare. We took turns pulling on the ropes and pushing the trolley up the slope. The pain in my leg was becoming unbearable as we struggled up the slope.

When we got to the top of the dike we found a heavy steel cable on each side, meant to keep cattle off of the road and vehicles on it. In the distance a building was on fire, lighting up the area with an orange glow as the white light of the flares lit up the meadow. Tracer bullets flashed across the meadows as deafening explosions of artillery and mortars and the screech of Nebelwerfers sent a cold fear up and down my spine. All this time we had to lift and pass the trolley, holding it level over the road, and down the other side.

While moving the trolley down the slope of the dike was easier then moving it up, we could not afford to have it damaged, and we were now exposed to enemy fire on this side of the dike. The scene was crazy as I saw men carrying the heavy assault boats, and falling to the ground as shells burst and machine guns swept the meadow. The ground was so soft that the wheels on the trolley were useless. It seemed as it took forever before we reached the scant shelter of the low dike. We were safe from machine gun fire as we lay on the cold, wet grass, but the tracer ammunition gave the horrifying impression that every bullet was aimed directly at you. However, the low dike provided little protection from artillery and none from mortars.

Peeking over the low dike, I saw Lieutenant Grunbaum from our brigade's engineer company standing in water almost to his knees as he calmly directed the launching of the boats. I saw boats making their way across the river carrying our signalers. I noticed that my friend Czesław Gajewnik, who was turned away from a boat full of signalers swam behind it as it made its way across the river. There was no let up in the enemy fire as I saw boats getting hit, and sinking.

The sky was turning from black to dark gray when a mortar bomb exploded near Lieutenant Grunbaum and I saw him fall. Some of his sappers carried him behind the low dike. It was not long before word was passed down to stop the crossing and that we were to withdraw behind the dike. We began moving back in small groups, or as individuals to the cover of the large dike.

Driel
September 24, 1944

When we returned to our old positions in Driel I glanced at my friend Boguniewicz, and our eyes met briefly. No words were needed. We were beyond hunger and exhaustion. All I knew, was that I had stared death in the face and was lucky to be alive. We had to get our radios back in action, and take stock of who from our company still remained in Driel. Lieutenants Wilk and Rakowski were with us, but many others were missing. We all hoped that they had safely made it across the river.

Among the British in Driel that morning we saw a high-ranking officer with a scarlet band on his hat. I later found out that this was the commander of XXX Corps, Lieutenant General Horrocks. He spoke with General Urquhart in Oosterbeek through our command radio. There was a lot of movement among the staff officers, and we had the impression that something was about to happen. We received news that our missing troops had landed in American-held territory, and that they were on the way to join us.

The enemy bombardment of Driel that day was the heaviest we had experienced. Our field hospital had been hit so many times that it had to be abandoned. The commander of my radio section, Lance Sergeant Sobieralski had stuffed his ears with whatever he could, as the constant shelling was driving him crazy. My sergeant was not unique. The four days under bombardment was unnerving many others.

We spent very little time on the air. Most communication was by telephone. The telephone platoon had no rest as they were constantly repairing the lines under the heavy bombardment. Very troubling was the fact that there was no word from the signalers who had managed to get across the river with a radio.

It was during the late afternoon when Lieutenant Wilk came with the new orders for that night. Brigade headquarters and the 2nd Battalion, along with all the remaining troops in Driel were to cross the river at the same spot where we had attempted to get across during the previous two nights. This was unhappy news. No one needed to be reminded of the way that the enemy covered the river and the meadows with fire. Their range cards and aiming stakes were obviously well-adjusted. The surprising news was that when our missing troops arrived, they would be crossing the river further west, under command of the 43rd Division. It was two hours before sunset when our company commander, Captain Burzawa and eight soldiers from the Signals Company joined us.

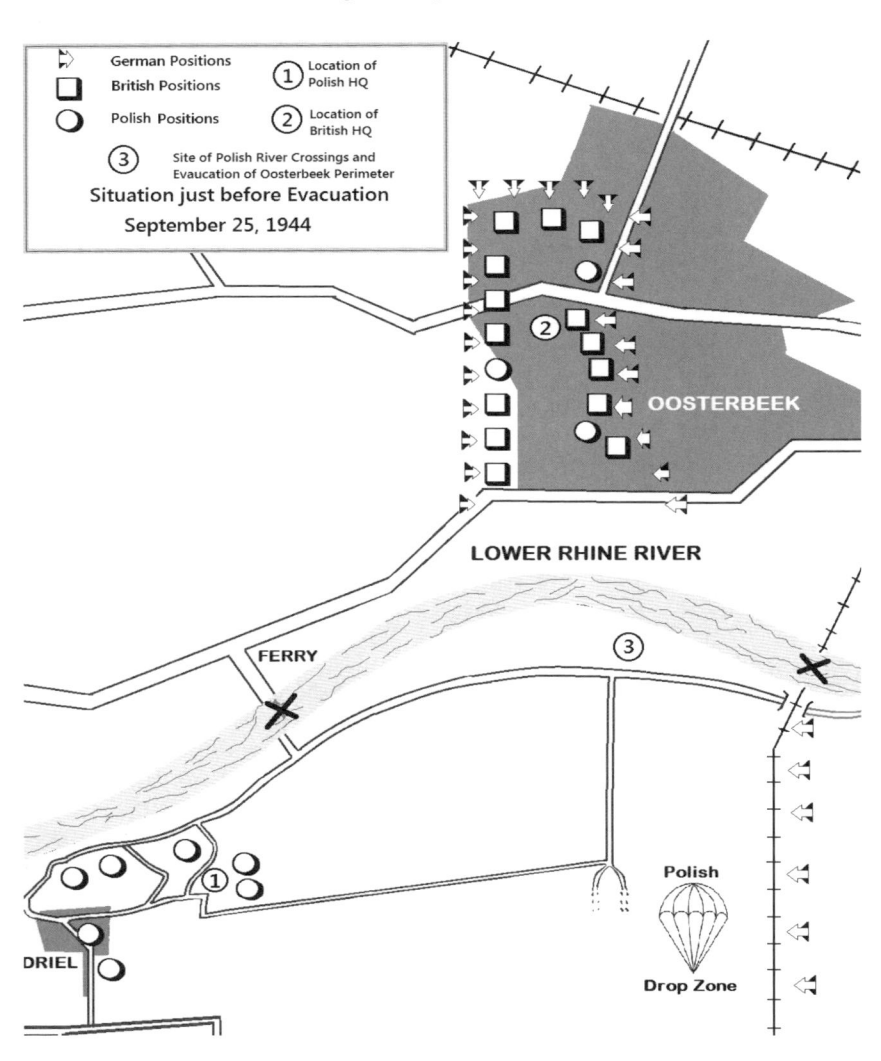

While the sappers went off to the crossing site at sundown to prepare the boats, we unhappily prepared to pack ourselves and our equipment and go to the crossing site. We were ready to move out to the crossing when we received a voice transmission from the sappers that there were British officers at the crossing site asking for our boats. The message was passed on to headquarters. Word came back that the 43rd Division had contacted headquarters directly, and that they needed our boats. Our part in the crossing was cancelled. We were greatly relieved, but were worried about our comrades who were to cross with the British.

Most of the artillery that night had fallen to the west. Lieutenant Wilk told us that the British crossing went very badly, and that our paratroopers did not take part. At dawn they began arriving in Driel.

The new arrivals in Driel did not escape the attention of the German artillery observers. The small village now felt downright crowded. As the new arrivals began to dig in, shells began falling. They did not have the experience that we had earned during our days in Driel, and often dug under trees where shells exploding in the branches killed several. This began a day of the worst shelling since we had landed.

The wounded were taken to our field hospital in Driel for treatment. Ambulances from our sea lift had moved ahead of the rest of the column, and with British ambulances began to evacuating the wounded. During the afternoon the hospital was heavily damaged, and two of our ambulance drivers were killed by the bombardment and their vehicle. After the remaining wounded were evacuated, the hospital was abandoned. The Germans even scored a direct hit on General Sosabowski's headquarters. That day three of our signalers were killed in Driel, and another wounded.

There was little we could do but hunker in our foxholes, and pensively wait for what the coming night would bring us.

There was a lot of radio traffic with the 1st Airborne Division's headquarters across the river, most of it from the British radio that had joined our net. Late that afternoon there was an officers call. Lieutenant Wilk told us that the British airborne troops will be withdrawn that night. Our brigade was to leave Driel the following morning. Wilk further told us that we should be packed and ready by sunrise, but also cautioned us that we should also be prepared should there be any change in the situation

General Sosabowski's Headquarters in Driel

Night brought heavy artillery fire landing on the edges of the British perimeter across the river, to accompany the evacuation of the paratroopers still alive north of the river. The Germans naturally retaliated, with shells landing in Driel, and on the side of the dike facing the river. Flares and machine guns joined the "symphony," and soon a hard and heavy rain began falling.

The night was almost surreal with the noise, the flashing lights and the soaking rain. Out of the darkness the first evacuees appeared. They looked like apparitions rather than the strong and strapping paratroopers who were going to bring an early end to this horrible war.

The sight of the soldiers who were evacuated from across the river was very depressing. They kept coming out of the black night looking like phantoms. Few had weapons. Some were bare-headed and without boots, others wore bandages that had not been changed for many days. Those who still had their uniforms were filthy and ragged. Many were almost naked, shivering and covered themselves with blankets, civilian clothes and even window curtains. The pouring rain did little to wash the filth from their and unshaven faces. When directed to a collection point where a mug of tea laced with rum and lorries to drive them to the rear awaited them, their red-rimmed eyes stared blankly and without a reaction. They were similar to the people I had seen in Siberia and on the shores of Persia, but these men had been reduced to this almost zombie-like state in less than ten days.

With dawn everything was strangely quiet. We made our final preparations to leave Driel. We were to march to the bridge in Nijmegen, where we would receive further orders. As our march would be through terrain held by the British, we would not have to send patrols ahead. The route was mostly along side roads to avoid Elst, which was still in enemy hands. Our military police were sent on ahead to direct our column.

It was nine in the morning when we left Driel. The roads were slick after the heavy rain from previous night, and the fields seas of mud. The Germans had woken up to the men marching across the flat terrain, so our withdrawal was punctuated by mortar fire. This kept our medical orderlies busy taking care of the wounded. Fortunately, none were from our company, but we now had to avoid the roads. Working our way through the orchards, and using the cover of ditches and depressions was very arduous .

Carrying or pushing or pulling everything we had with us, we were all wet from the night before. Ironically, there was not a cloud in the sky, and the sun was strong. Our layers of clothing became very uncomfortable under the weight of our loads, and our wool uniforms made us perspire heavily. Thankfully, there were still Dutch villagers who gave us what they could, mostly water which we drank vigorously, apples, or sometimes fresh milk.

We arrived at Nijmegen during the afternoon exhausted and near the end of our strength. The American paratroopers gave us a very warm welcome, patting us on the back and giving us the "thumbs up." A surprising number were of Polish origin who struck up conversations with us.

At the southern end of the bridge a column of lorries from the 1st Airborne Division's seaborne lift was waiting for us. Within an hour we were all loaded onboard, and started off in the direction of Grave. The lorries then turned east and we saw Dakotas taking off and landing. Our lorries stopped at meadow which had been turned into a make-shift airfield. There we saw the British paratroopers who had been evacuated from Oosterbeek climbing into Dakotas, which took off as soon as they were full. We sighed with relief. We were going back to England for a well-deserved rest.

There was an officers call, and when they returned to our lorries they climbed aboard with disappointment written all over their faces. The brigade was sent to the southwest and took over a sector close to Neerloon on the highway that ran north from Eindhoven.

The British 1st Airborne Division's seaborne lift replaced signals equipment and also turned over some Jeeps. We received new orders, and moved on until we reached the village of Neerloon. Here the brigade occupied a defensive sector, providing security for an airfield. Due to the nature of our of our defensive posture, the brigade relied on the telephone net and messengers for communications.

We now had to take roll, and try to see who was with us and who was missing, and to pray that they would somehow return to the company, and were not wounded, in the hands of the Germans, or dead. Lieutenant Relidzinski and Cadet Officer Niebieszczanski were the first to return to the brigade from across the river. Of the eight signalers who had crossed the river to Oosterbeek with Lieutenant Relidzinski, only Cadet Officers Niebieszczanski and Blazejewski had rejoined the brigade in Driel. Of the eight signalers who went into action by glider, only Cadet Pajak came back from Oosterbeek. He and Blazejewski were both wounded during the evacuation, and were taken by the British to their field hospital in Nijmegen. Cadet Niebieszczanski had been wounded by a shell fragment, but chose to be treated by the brigade's medical services, and remained.

Relidzinski reported that the radio they had brought across the river to Oosterbeek had been damaged beyond repair, and that he and his signalers were assigned to hold a position as infantry. It was further reported that all of the wounded were evacuated to hospitals that were in German hands, and it was assumed that all were now prisoners. When I asked about Czeslaw Gajewnik, he said that he was last seen on the northern bank at the edge of the river, and knew nothing further.

The following day our seaborne lift arrived, bringing six men and spare equipment to our company. A dark cloud hovered over the 1st Polish Independent Parachute Brigade. We had made great efforts to make the operation a success. From the frustrations caused by the delays due to weather, and our efforts to get across the river to help our British comrades in arms, the failure weighed very heavily on us. A highly trained and elite British division had been destroyed, and our brigade itself suffered twenty-five per-cent casualties.

I was depressed by all of this, and like many in the brigade, could only go through the motions of soldiering. We were in a long corridor with the enemy on both sides, and a few men were wounded by enemy artillery. As matters could not become any worse, on the morning October 4, it was announced that the Polish underground army in Warsaw had capitulated after over two month of vicious combat. For once the German propaganda machine had stated the truth with a radio broadcast that announced, that " after weeks of fierce fighting, which has led to the almost total destruction of the city, the remaining rebels, deserted by all their Allies, have given up and surrendered."

A few days later, the brigade received orders that we were to withdraw to Belgium, and were to return to Britain. The same night that these orders were received a British patrol mistook our soldiers for Germans, and opened fire. Two paratroopers were killed, and two were wounded. One of them would die in hospital a few days later.

The following morning, October 7, we packed up our equipment and boarded lorries for Louvain, where we spent the night in former German barracks, and slept in their old bunks. The next morning we were taken to Brussels airport, and waited in vain for the airplanes which were supposed to fly us to England. That night we spent sleeping on the bare concrete hanger floors, without mattresses, or straw, or even blankets. After spending another day at the airfield, we were told that there were no planes for us. It was after midnight when we boarded lorries for the port of Ostend.

After an afternoon meal in Ostend, we were loaded on landing ships. That night we began the final stage of our journey home. After a night in the smoky and stuffy berthing spaces, we landed at London's Tilbury Port on October 12. We and our equipment were loaded onto trains, and rode back to the same billets that we left twenty-two days before. The local people greeted us with great warmth and joy, which we had not really expected from the normally reserved English. Among those greeting us were the men from the brigade's second seaborne lift. They had all ready been back from Normandy for several days, having seen no action.

Though we were thoroughly relieved to be back in Britain, we paratroopers were far from happy. Our casualties were severe, and over a third of our Polish parachute signals company was either dead, missing, in hospital, or behind barbed wire. Warsaw had fallen, abandoned by the Allies, and our future was uncertain. It was difficult to return to our old billets with so many no longer here.

Part 6

Uffington, England
October 10, 1944

The Brigade went about assimilating replacements, and rebuilding itself. Many of the new soldiers who were joining our ranks were youngsters, who had been conscripted into the German forces, and then surrendered to the Allies. The Our signals company moved to new billets in Uffington. But somehow things just did not feel right. We were all more than a bit surprised when the British Parachute Training School at Ringway stopped training Polish paratroopers in November, 944..

With the beginning of winter, shocking news flashed through the our ranks like lightning - General Sosabowski was leaving the 1st Polish Independent Parachute Brigade. There was no explanation.

Like any good soldier, I avoided unnecessary contact with officers not in my unit. General Sosabowski was a man with a prickly character, and was feared and often disliked by senior officers. Frequently impatient, he would give a fiery dressing down to anyone who did not carry out orders 101%, or otherwise offended his will. The only ones who were exempted from this treatment were young junior officers or cadets who were quick thinkers, and showed no fear of the general.

Frequently a martinet, woe be it to the common soldier who wore his battledress blouse with the top button open and a necktie underneath in an attempt to attract the ladies. But despite his prickly personality, the soldiers trusted him, and knew instinctively that though we were in a position as paratroopers to take heavy losses, he would not sell our lives cheaply. We really believed that he would lead us to Warsaw.

Rumors flew that there was a hunger strike in some units. The rumors were followed by the general joining the soldiers in their mess halls, and admonished them that military discipline must and will be maintained. Christmas Eve was his last day as our commander.

With the new year, we were spending more time in the field on maneuvers with other British airborne units. All along, my leg was feeling worse, and at times I was in considerable pain. I went to Lieutenant Wilk, and he put me on medical leave, and on May 15, I found myself in a hospital bed. At first there were fears that the leg would be amputated, but after extensive surgery, it was saved.

Upon leaving hospital, I was told that I would need recuperation, and I was sent to a convalescent home on the island of Anglesey in northern Wales. There, I spent the most pleasant weeks of my military career with plentiful food and light duty.

When I was pronounced fit for duty, I unfortunately I found myself without my army home. The entire 1st Polish Independent Parachute Brigade had been moved to Germany as part of the British Army of the Rhine for occupation duty. I was sent to the Brigade Depot in Scotland

Nothing was happening there, other than military discipline inflicted on us by superiors who seemed to resent the fact that they were sent there instead of Arnhem. After several weeks of my making a pain in the neck of myself, I finally was transferred to the Parachute Brigade in Germany. Though I was again with my comrades, my best friend Tadek Boguniewicz had been discharged, and was attending university in England.

Gerde, Germany

The brigade's soldiers had gone back to the battlefields of Driel and Oosterbeek to salvage equipment. They were surprised to find that the people of Driel had maintained and decorated the graves of our fallen comrades with red and white flowers. By the second anniversary, the Dutch villagers, led by that Baltussen family, had erected a monument and hosted memorial ceremonies on the anniversary of the battle. One bit of sad news was that my good friend, Czeslaw Gajewnik's body had been found along with that of a Canadian sapper in a shot-up boat downstream from Driel.

We lived in the small town of Gerde. We were some 5 km away from Bersenbruck. We continued to train for combat, but I soon found myself in the signal company's headquarters. By that time I had been promoted to lance sergeant, and with a colleague From our headquarters company, found lodgings in a private home. German dwellings had been requisitioned for use by the occupation troops. The common soldiers continued to be billeted in barracks, but majority of the NCOs were billeted in private houses

In the signals company we had twenty-odd soldiers who had been conscripted by the Wehrmacht, and naturally they spoke German well. They started taking up with the German girls. There was a shortage of everything in Germany, and for some cigarettes or food you could get the prettiest German girls.

There were Polish women, both former laborers and military veterans of the Warsaw Uprising attached to the other Polish units occupying Germany. They helped out in the various brigade offices. The ones attached to the Signals Company worked the switchboards in the brigade's telephone exchanges. Others were worked in the canteens.

Our brigade offered technical and trade courses which, it was hoped, would provide us a start in civilian life, but I never signed up. In the meantime the opportunity to return to Poland now became a possibility. The Government of the "Polish Peoples Republic" called for the Poles remaining in the west to return. The subject of returning, or remaining in the west was hotly debated among the soldiers. This serious question tormented us without mercy, and left us little peace. I was no different from the majority in the brigade, we all had our futures to think about, and the question divided us in a manner unlike any other matter had before.

A Final Review For Our Comrades Who Decided To Return To Poland

The one item that was staring all of us in the face was the fact that Poland was no longer free, and our motherland was strongly in the clutches of Soviet communism. Our beloved nation was being ruled by Moscow by a mere handful of Stalin's Polish flunkies. Having had an inside view of Soviet communism, it was not difficult for me to arrive at my decision. I would remain in the west and rebuild my life. It had been more than seven years since I had seen my home in Poland, and the decision was not taken lightly. The hardest part was the fact that I had no information about how my family had lived through the tragic years of war and occupation, or even the fact if any of them were even alive.

And in April, 1947, the Polish Army was relieved of its occupation duties within the ranks of the British Army of the Rhine. We started packing for a return to Britain, and demobilization. Taking my first steps back into civilian life I realized that I had closed a major chapter of my life. Though filled with deprivation and suffered, I also had my share of unbelievable and fascinating events, which in some ways were rewarding despite the misery. The events of those days were coming back to me with increasing frequency. Most of the time I was unable to believe that I had actually survived those many hardships. But, I would stop and think of the many others who had endured even more tragic circumstances than mine. After all, hundreds of thousands of families with children from Poland's eastern Borderlands were inhumanly deported to the various corners of the "Soviet paradise."

A Not So Merry Christmas In Germany, 1946.
Might There Be A Better a Better New Year?

Part 8

England
June 20, 1947

We returned to England as soldiers, but without military duties. We were assigned to the newly formed Polish Resettlement Corps (*Polski Korpus Przysposobienia i Rozmieszczenia*). The Corps served as a holding unit for members of the Polish Armed Forces who did not wish to return to a Communist Poland. We were subject to British military discipline and military law. Here we were we attended English language lessons, and either given training in trades or loaned to private contractors. It was meant to ease our transition from military into civilian life until we could begin our lives in Britain.

This was about all the help we received for our new "beginnings". We were interviewed by representatives of the Minister of Labour and National Service. I managed to stretch out my interviews as long as possible, with the hope that I would be able to get the best circumstances of employment. In the end, I decided to seek employment in the textile industry.

I was now no longer a soldier. I was also received the handsome sum of 30 pounds as a "war gratuity". This was my reward for five years of military service, and my participation in the Battle of Arnhem. I also received a suit of civilian clothing and a railroad ticket to Manchester, my place of employment.

I checked into a hostel run by a former Polish soldier. He invited me for a tea of tea, and told me to leave my suitcases in an empty room. When I returned for them, they were gone, along with my civilian suit and everything else I had accumulated from the time since I had arrived in Persia five years before. Not only was I angry, but I was also stunned because the hotel where this happened was run by Poles, and that one of my countrymen would give me such an introduction to civilian life. I had no savings, nor a trade, and did not even have a toothbrush. This event left me embittered, and left me with not very promising prospects for my future.

Rochdale
February 10, 1948

The factory that hired me was one among many in Rochdale, a half-hour by bus from Manchester. The bus would arrived early in the morning, while it was still dark. I was lucky that an old comrade from the Signal company, Jozef Kutereba was working with me. I started at the bottom as a common laborer, but I was supposed to learn everything about the factory with becoming a textile machinery repairman as my final objective.

The machines in the factory were very old, and some of them even dated back to Victorian times. There was always something going wrong with them. From the very first moment I realized that the mechanic from whom I was supposed to teach me everything constantly had his hands full due to the age of the machines. There were only women working in the mill, and I felt strange about working in such a circumstance, but thought that maybe with time I might have become more comfortable in such an atmosphere. My wages were very low - three pounds and 10 shillings, and I realized that despite my best efforts to improve myself and my position, I could not count on my conditions changing for the better.

I could not tell you what was the worst aspect of working in the mill. The noise was incredible, the banging of the looms, the clattering spindles and the whirring spinning machines dulled all of the senses and made thought difficult. And then there was the dust. The ancient ventilators were usually clogged, if they worked at all. The air was always full of something that had to do with cotton. Whether it was wisps of lint drifting in the currents generated by the hot machinery, or microscopic dust that clogged the nose, parched the throat or inhaled deep into the lungs. The heat and humidity were insufferable. Given the circumstances, my depression began to deepen.

The war and managed to inflict unbelievable transformations, injuries and losses among our loved ones I had known many bitter experiences during the war, and these did not stop with the end of the fighting. I was demobilized with six years of army service behind me, without the slightest bit of information about my family. I found myself without any skills necessary for a trade - how, under those circumstances how could I have possibly learned one? My ability to speak English was far from the best . . . .

I felt disappointment with everything, and was deeply

depressed. Would this end with my mental collapse? I had not stopped thinking about Jozefa for a single moment. But what of the difficulty in which I found myself now? I asked myself the difficult question, would I break? This was my sad reality at that time.

My hospital stay due to my injured leg, and later occupation duties in Germany did not help my situation in the least bit. Then, in June, 1948, I received a long letter from Jozefa; it began with the words: Dear Bolek; I have been searching my soul . . .

I had to struggle to read the rest of the letter. I almost felt her sadness as she told of pressure from her family, and with no sign that I would ever be discharged from the army, she was left with a spiritual dilemma. She then explained, that with a heavy heart, that she would have to let me go, and had decided that she would marry a Polish sailor she had met. She closed the letter asking if she had made the right decision? And that it was now in the hands of God. I felt cornered, and had to do something so I would not suffer a complete collapse.

This was my life until the summer of 1948. I had managed in the meantime to again locate my sister's family. I last saw them in Persia, after they had survived living in the Archangel Oblast, and found them living in Coventry, in the English Midlands. My brother-in-law had served in Italy with General Anders' Polish IInd Corps, while my sister and niece were among the Polish civilians who spent the war in refugee camps set up in India. It did not take me long to make my decision. I bought a suitcase to replace that stolen. It was small, as it was all I could afford, and besides, I did not have very much to put in it.

Coventry
1948

So I waved the textile factory, and all the rest of Rochdale good riddance, and bought a train ticket for Coventry. The city was an important industrial city near Birmingham, and some 100 miles northwest of London. Coventry achieved unwanted fame as a result of the vicious bombing in 1940 that destroyed the entire city center. Finding a place to live was very difficult, but my brother-in-law, as a result of being a veteran, managed to get accommodations on the edge of the city. This was a former army barracks which was quickly built as a temporary facility. I moved into the tiny corner that the three of them had been allotted. Despite the fact that it was very crowded, there was warmth and happiness there as we were together again as family.

My sister took care of the cooking and cleaning, while the three of us went to work. My niece Irene, having learned excellent English, was almost immediately hired by GEC, a large electronics corporation, and she thrived there. It it seemed that the majority of people who had spent the war in the Valivade Refugee Camp near Kolhapur, in central India, had recovered admirably from their time in the Soviet Union. Even better, it seemed that the majority who came from Valivade arrived in Britain speaking excellent English. My brother-in-law's English was halting, and as a result he had to settle for employment as a construction laborer.

I had already mentioned that we were getting along financially, but our living arrangements were very tight. I did want to get married and have a family, and immediately upon arriving I thought that Coventry suited me well. It was well known that central England always enjoyed better weather than hazy and humid Manchester, and in my own mind, I anticipated that this would be the place where I could acquire skills with which I could build a professional career.

It had been three years since the end of the war, people were constantly moving into more comfortable, or affordable living places, or moving to other towns in search of better conditions. For we Poles, decimated and scattered throughout the world, this was one of the most difficult periods in our history.

It so happened that shortly after I had moved to Coventry, an entire family which had lived in Volhynia moved in nearby. It was indeed a rare event that an entire family from Eastern Poland had survived the war, and managed to find each other and be together again. What was even more unusual was that they had come from nearby Bortnica where my sister had lived with her husband. They were the Wypijewski family, whom I had known from before the war. I did not know them well, and knew their children even less. This was one of the reasons I became so pleasantly enchanted was when one day, during some occasion, when I met Maria and Krysytna Wypijewski. Who would have thought that with the passage of a few years two little girls to whom I had never paid the slightest bit of attention to would bloom into appealing and very attractive young ladies.

At any rate, as a soldier whose boots had tread on several continents, and I had known a number of ladies of different nationalities. But, I have to state that none could hold a candle to Polish women. From that moment I saw the sisters regularly.

However, I was now faced with a serious problem; which of the two should be focus of my attention. Krysia, slightly younger of the two, intrigued me. She was always the joyful, merry and laughing scatterbrain, whom it was impossible not to like. Mary turned out to be more mature and quite serious for her age.

But much to my chagrin, the visions of my future which happily danced around in my head changed drastically. I discovered that Mrs. Wypijewski insisted that the eldest daughter was to be first to enter the bonds of holy matrimony. With this "dictum" established, the only thing left for me to do at any rate, was to start courting Maria.

At this point I must digress. During the entire time that I was in the parachute brigade, and especially during the battle of

Arnhem, I had formed a very close friendship with and Tadeusz Boguniewicz. Unfortunately, after the war we had to go down our separate roads toward our futures. Tadeusz went on to study at the Polish University College in London. Despite this, we continue to remain in touch, and after I had settled in Coventry I invited him to visit me.

However, when he did visit, I didn't have the slightest idea about what events would set in motion, and Krystyna had his head spinning. Tadek fell in love with Krystyna, as the saying goes, "at first sight." I do not know whether Krystyna reciprocated his feelings, but my friend returned to his studies in London an entirely different person, but dozens of letters filled with poetic yearnings found their way to the Wypijewski's mail box during the following months. The "new" Tadeusz Boguniewicz, unlike his normal quiet and subdued self, turned out to be not quite so reserved, but possessed a emotional and romantic streak which appeared on paper that never manifested itself in person.

The first years after the war were difficult for we Poles who had decided not to return to a communist Poland. The majority of us lacked any employment skills or qualifications. I remember from my time in the army, only a very few of my comrades were involved in careers or had technical training, and they were usually a few years older than the rest of us. Another obstacle was the fact that the majority of us had not mastered the English language fluently. The former Polish soldiers were usually steered toward work which nobody else wanted, poorly paid, dirty and physically demanding.

One has to remember, that Great Britain emerged from the war having expended its treasury, and stripped of much of its empire. A multitude of recently demobilized soldiers and sailors were seeking work in a nation geared to war production that was no longer needed. The Polish soldiers were an added problem in this situation, and the majority of the British would have been very glad to see us return to Poland. All of us remember Ernest Bevin, the British Minister of Labour during the war, and then the Secretary of State for Foreign Affairs afterwards, and his declaration that we should return to Poland now that it was "FREE". Despite the gratitude of those of us who were delivered by the British from hands of the Soviets, we now felt exploited, and to be discarded as quickly as possible as they no longer had any use for us.

With all of this going, I made the decision that I had to start my own family. I was then thirty-one years old, and I realized that there was no real reason to delay. As I had mentioned before, I met a Polish girl, Maria Wypijewska from Bortnica. We were like minded, and communicated well. We both had similar goals in life, and we each had a "true love" in our past. However, we were both over 25 years-old. Our "green years" were behind us, lost forever because of the war. Though one could never retrieve the past, as I look back, I realize that we both determined to get what education we could, find meaningful work, and start a family. Needless to say, our road was difficult, but we never entertained of giving up.

In 1949, Maria's parents took all of their savings and with a mortgage from the bank bought a small row-house at No. 5, Duke Street. This small and very modest house was to be home to the parents, and five fully grown children. Whenever I had dropped in, I found a warm, hospitable and cheerful ambiance. All one has to remember, that there were three wonderful young women lived there, and what other atmosphere could one expect? Especially as there were always young men and women there, and I always had my eye on the girls. My big problem, was how to stay ahead of "them" - those other Polish guys who were also observing the landscape. Among them was my good friend, Tadeusz Boguniewicz and the accordion playing Jerzy Rytwiński.

Those looking to propose marriage do not necessarily have to choose a special place to ask the big question. It might happen at home, on a walk, at the theater and so forth. I proposed to Maria on the bus while we were returning home from a night at the cinema.

Naturally, it took me quite a long time to decide to propose to Maria. According to my future mother in law, it was all taking too long, and resulted in my receiving letters and some not very subtle suggestions from her. However, I always believed that important life decisions should not be taken lightly. Unlike Maria's mother, all of our friends knew that we would eventually go to the altar.

Despite the fact that Maria's mother was pressing me, I felt sorry because of the pressure on Maria was even heavier on her. I understood, and let her know that I understood, and everything went smoothly. The wedding took place on January 20, 1951, in a small church, Christ the King, on the edge of Coventry. Due to the circumstances, the ceremony and celebration were humble, and the reception was limited to family and a few close friends at No. 5 Duke Street. Despite the modesty of events, I remain grateful to my mother-in-law, who greeted us in the traditional manner at their threshold with bread and salt. The times were not easy for anyone, and she did everything she could to make our day memorable.

But, after the wedding, there was no way that our financial situation would permit we newlyweds to stand up on our own two feet. As a result, I moved in with the in-laws at their home. We were living in a tiny room upstairs. A year later, there were now three of us when Maria gave birth to our son Janusz, who from the beginning my wife called "Januszek", or "Johnny Boy".

I certainly did not want to remain a factory worker and with the hope to of gaining a more meaningful position, I began a three-year engineering course at the Coventry Technical College. I worked the night shift at Coventry Swaging Company, a manufacturer of screws and other metal fasteners. I would come home from work and sleep, wake up, complete my assignments, and then attend classes at the college during the evening, and then work my shift.

This was my life for three years until I received my certification as a technician. I was immediately hired by Massey-Ferguson, a manufacturer of agricultural machinery as a draftsman.

With the three of us crammed into the tiny room, something had to be done. On one hand, I was satisfied. Though I was not making much money, I had steady employment, and I knew that I was a good worker and well respected by my employer. I was counting eventual promotions and pay raises. This was the situation when I began patiently looking around for a place that would become our "nest". In the first years after the war, there was a terrible housing shortage. After all, industry was fully mobilized to producing the tools of war, while the German bombers and V-weapons destroyed housing. There was practically nothing for sale, and should something turn up, the financial commitment was huge. At any rate, I would have to borrow the money for a down payment for a mortgage from a building society.

After we had looked at a number of houses, we settled on a house at 22 Lammas Road, a quiet side street just a 15 minute walk to No. 5 Duke Street, and just far enough away from the in-laws. Maria was very happy because she finally found herself being the "lady of the house" in her very own "dump". And so, we began setting up a household. Our relations with the in-laws improved immediately as we were no longer on top of each other, and Maria was in her own element where she was able to happily entertain guests.

Here, on September 26, 1954 our second son was born. Roman was blond and very different physiognomy from Janusz, so I jokingly told Maria that he bore a startling resemblance to our blond-haired milkman. Our daughter Barbara completed the trio of our progeny. She was born on December 15, 1955 at home, and not in a hospital. During those years in Britain, woman was allowed to have her first child delivered in hospital. I found myself in a rather tricky situation when Maria woke me up way past midnight with the words, "The baby is on the way". Where was I going to find a midwife at this time of night? I was running around, here, there and everywhere on a bicycle, as we did not have either a telephone or an automobile. At seemingly the last moment, I finally found a midwife and brought her home as Maria was going into labor. So, in four years we managed to expand our family from two to five.

In 1958, after a long and hard period of decision making, weighing both the benefits and disadvantages, we decided to emigrate. There was one more major decision to make - which country would we move to? We finally decided on Canada. Why did we choose Canada? Krystyna longed for her beloved boyfriend, Jerzy Rytwiński. Jerzy had gone to far-away Canada earlier, "for bread", and in the meantime managed to feel very much at home there. Krystyna followed him to Toronto, and they married during September, 1952. Krystyna's bold move was one of the deciding factors in making our decision.

At the same time an apparently much more important factor was thrown into the equation - the Suez Crisis. In 1956, Britain and France found themselves in conflict with Egypt over the Suez Canal. The First Secretary of the Communist Party of the Soviet Union, Nikita Khrushchev threatened Britain with nuclear war. The people of Britain were shaken and appalled by the prospect, and, as a result, thousand of Britons emigrated to the Commonwealth nations, or the colonies, most of which were still in existence at that time.

In summary, I was not a British citizen, and considered a "foreigner." We were living on an overpopulated island, and I questioned what my future would be in case of a conflict? Would I even be able to hold a job? Massey Ferguson had a design bureau and a factory in Canada, and I thought that might another factor in emigrating there. By that time, I had received several promotions, and was now working in the design section. I went and had a chat with Mr. Chambers, the chief engineer, asking for a transfer. To my surprise, he responded positively. Things were beginning to look up!

Canada
1958

Early in the spring of 1958 I made the ocean voyage to Canada. I was alone when I landed in Quebec, a province redolent with the pleasant scent of forests. In the event that something did not go right in Canada, I was assured that I would be able to return to my job in Britain. My contract in Canada was for six months. With the family still at home, I could establish myself with my "new" employer and get used to the rhythms of life in a new country without distraction, and without exposing my family to either risk, discomfort, or disappointment.

It turned out that my fear was justified. Canada was going through a recession, and shortly before I arrived at Massey Ferguson many employees had been dismissed. My situation was difficult and I was going to find a job elsewhere, and then go back to England. In the end, thanks to the intervention of Mr. Chambers in Coventry, Massey Ferguson did offer me a job, and I brought the family over seven months later.

Draftsman

Canada turned out to be no less difficult for the newly arrived immigrant than England. The nation had nothing similar to the social safety net in Britain, with its minimal medical and welfare benefits. Finding work was very difficult, and every new arrival had only to rely on their own physical and mental strengths and skills. In case of serious illness, complete financial ruin was probable. On the other hand, with some luck and a lot of skill and energy, one could eventually stand tall on his own two feet in a relatively short time.

And so, despite some difficult beginnings, we managed, although my wife repeatedly let me know that she wanted to return to England during the first years. In the end we were able to adjust to life in Canada. One after the other, the years passed. Both of us worked (or, I should say toiled) as we coped with mortgage payments on our house, and to ensure that we had the financial resources to provide an education for our children. Now, when I look back on those difficult times, I remember a lot of hard work, and years of scraping and saving.

I retired in 1984. Despite the fact that there were many Polish veterans in the Toronto area, I was mainly involved in keeping a roof over my family's head and putting food on the table. As the years passed, and we assimilated into the folds of the Canadian middle class, I established friendship with the local Polish veterans, which included a number who had served as paratroopers. I joined the Polish Airborne Forces Association (*Związek Polskich Spadochroniarzy*, or the "ZPS"), which was founded during the Parachute Brigade's days in Germany. The association published a quarterly bulletin *Spadochron* (Parachute). This publication was eagerly awaited, and helped former brigade members to stay in touch, despite the fact that we were scattered across the four corners of the planet. For decades this bulletin provided us with information about how and what our comrades were doing, along with news of their families. This was especially important to our comrades who had returned to our motherland, and had found themselves behind the Iron Curtain. *Spadochron* also published the announcements of our brigade's reunions and details of forthcoming pilgrimages to Holland, in whose soil we left many of our comrades to their eternal rest.

The first time I returned to the Dutch battlefield where our Polish Parachute Brigade fought was in 1989. It was the 45th anniversary of the battle, and thousands of British and Polish veterans where there for the event. I was in the company of my good friend and comrade, Tadeusz Boguniewicz. The hospitality that greeted us veterans was beyond description. We were a bit amazed that the Dutch people, who had suffered so much under German occupation, were liberated by paratroopers in just a little over a week. With the Allied defeat and withdrawal, the Dutch people were left with destruction, death, and starvation for another nine months before final defeat of Germany.

1989 - Tadek Boguniewicz And I Return To The Battlefield In Holland For The First Time.
My Hosts, The Van Solen Family Offered Their Friendship And Hospitality Twelve Times Over The Following Twenty Five Years.

    I continued to participate in these battlefield reunions, twelve times, the last being in 2014. In addition to their invitations to attend the anniversary commemorations the Dutch organizing committee partially paid some expenses for a veteran and his companion. In the case of those veterans who were living in Poland, they covered the entire cost. The van Soelens also received my family members as if they were my own. Over the years my daughter Barbara, my son John, and my grandson Erik made the pilgrimage with me.
    I had extended an open invitation to the van Soelens to visit us in Canada. I was gratified when they accepted, and visited for the first time in 1992. They had a delightful visit, and loved Canada, especially the area around Toronto. I arranged for them to visit the Canadian wilderness, and we spent several days camping and fishing in Algonquin Provincial Park.
    The van Soelens enjoyed their visit so much that returned a second time. On this occasion they planned to tour in a rented caravan, and visited large sections of eastern Canada. They returned to Holland with wonderful memories, many of which they preserved in color slides.

We who lived in freedom watched as step by step, the people of Poland managed to painfully and slowly regained their own freedom. In 1989 elections were finally held, and after 50 years Poland again became a democracy. Veterans from around the world were invited to a long-denied victory parade in Warsaw. I was among the former Polish paratroopers who were again united with their comrades in their motherland. We were also welcomed by currently serving Polish paratroopers in Krakow, whose unit had been officially renamed the 6th General Sosabowski Air Assault Brigade.

But our reunion was bittersweet. We, of course, were very greatly interested in the lives of the veterans who had returned to Poland. Many of our comrades who had returned to a Soviet-dominated Poland had suffered repression or imprisonment. Most were unable to gain suitable employment from the communists simply due to the fact that they had fought in the ranks of the western Allies. They had lived in poverty, but their spirits were often bowed but rarely broken.

General Sosabowski passed away in 1967 at the age of 75. Living on the wages of a common laborer, he had published two memoirs before his demise. Both of these books reflected a bitterness that was never fully explained. In 1970, A BRIDGE TOO FAR was published. The book, by Cornelius Ryan, was a worldwide best seller and was later turned into a major motion picture. Rather than portraying Operation Market Garden as a "partial victory," the book showed that the heroic endeavor was in fact a disaster.

For once the participation of the 1st Polish Independent Parachute Brigade was accorded more than a brief mention. General Sosabowski was shown to be opposed to the operation, and sadly his predictions of a catastrophe came to pass. As years passed, more books were published and the courageous epic of the British Airborne at Arnhem was continually re-examined.

Fate was not kind to we Polish paratroopers, though we all knew, from the lowliest paratrooper to General Sosabowski himself – we had spared no effort to come to the aid of our British comrades on the other side of the river. It is easy to imagine the shock and pain that we suffered when it was revealed that General Browning had Sosabowski removed from command with slanderous allegations that the Polish general and his troops were "quite incapable of appreciating the urgent nature of this operation."

Having our brigade used as a scapegoat for the British command's unrealistic planning and expectations of the entire Market Garden operation. A Dutch journalist produced a television documentary about our brigade and General Sosabowski. It exposed the scandal, and Prince Bernhard, commander of the Dutch forces during the war, was interviewed. He lamented the treatment of Sosabowski and called for recognition of the Polish soldiers who had played a great part in the liberation of the Netherlands.

Prince Bernhard passed away shortly afterward. His remarks were treated as a deathbed wish, and the Dutch Ministry of Defense began studying the matter. Cora Baltussen, who had also been working long and hard for the Poles to be recognized, was kept apprised of the situation. When asked for comment, she dismissed the journalist with the words, "seeing is believing." Cora did not live to see Queen Beatrix pin the medal and the sash of the Order of William, the highest Dutch military decoration, on the colors of Poland's 6th General Sosabowski Air Assault Brigade.

Queen Beatrix Decorates The Colors Of The General Sosabowski Air Assault Brigade

All the surviving Polish veterans of the Battle of Arnhem were invited to The Hague to witness that event in May 2006. I, along with the other veterans who were able to travel, was flown to the Netherlands to attend the ceremonies. The occasion again brought us to Driel to gather around our monument in the center off the village, and to pay our respects to Cora at her grave in the nearby churchyard.

It was an embarrassment of riches when a pair of distinguished British officers proposed erecting a monument to General Sosabowski in Driel. Sir Brian Urquhart was General Browning's intelligence officer, and was dismissed because of his dire warnings about the strength of the German forces in the Arnhem area. After the war he joined the staff of the United Nations, and held a number of critical positions. He accepted the 1988 Nobel Peace Prize in the name of the UN Peacekeeping Forces. Together with Tony Hibbert, a staff officer with the 1st Airborne Division, he led the monument initiative. The beauty of this monument is the fact that the donations were collected by and from veterans of the British airborne forces.

FINAL REFLECTIONS

Despite the fact that I had spent three-quarters of my life in the English-speaking world, I never felt fully integrated into Anglo-Saxon society. From my own experience, I found that the English people the English people held some sort of prejudice toward we foreigners. Upon my demobilization from the army in 1947 after five years of service, I was given a gratuity of £30 and a set of civilian clothing. My entry into civilian life did not at all look promising. I truly felt that I was at a crossroads. After serious deliberation, I ruled out returning to Poland. The reasons were obvious: after the Yalta Conference, Poland, as well as much of central and eastern Europe, was in the hands of the brutal Soviet empire. I wonder if the souls of Roosevelt and Churchill feel any regret for the decades of captivity and Cold War that followed.

Now, as I continue my solitary journey toward the "setting sun," the bright memories of my life as a child and as a young man are not forgotten. Yes, there are also memories that are dark, but fortunately, never without hope. There are other events that I recall as being just plain stupid, but these memories will never be erased.

Looking back at all of this today, and reflecting on all that I have experienced during my life, I am unable to resist saying that there exists a generation gap between those of us who were born in a free and independent Poland – of which I count myself as a member – as opposed to our children, and even more so our grandchildren, who are now part of the generations born and raised in the western world. This is a source of serious concern and worries to me, though I realize (and am resigned to this) as I know that this trend in consumption will continue in generations after mine. As a father, and as a grandfather, I've tried to inculcate these values, which have always been very dear to me, to my children and grandchildren. As I observe the lives of the current generation, which had been raised in an atmosphere of unbridled materialism and liberalism, and often lacking fundamental moral and spiritual values, I find it difficult to imagine a proper functioning contemporary world.

There is among the voluminous literature about the Second World War, novels, reminiscences, and memoirs concerning the fate of millions of Poles during those years. They concern those who dedicated their lives to defend their motherland in 1939, those murdered or sent to death camps, or deported as slave labor in Germany. The litany of horror continues in the part of Poland invaded by the Soviets. Polish citizens were tragically torn from their homes during 1940 and 1941, and deported to labor camps in the far northern and far eastern regions of the Soviet Union. Many of those who remained were bestially murdered by the nationalists of UPA (*Ukrayins'ka Povstans'ka Armiya*, the Ukrainian Insurgent Army) who wanted to see Volhynia and Eastern Galicia "ethnically cleansed" of Poles, Jews, or anybody else who did not share their maniacal vision of a Ukrainian nation.

I have had no success, until now, in discovering the memoirs of young Poles from the borderlands who were forcibly conscripted by the Red Army, given the most rudimentary training, and when the Germans turned on their Soviet "friend," were sent immediately to the front. I mention this because I was one of them. Wearing the red star on my forage cap, I with other young Poles swore allegiance to Stalin's constitution, and having to declare that with our "free will", to give our lives for our Soviet paradise. The slightest hint of dissatisfaction of the whims or wisdom of the Soviet system would lead to a death sentence, or at least years of deportation to a faraway gulag. That I managed to survive those hardships in one piece was through a combination of good luck, fortuitous circumstance, and the hand of God on my shoulder.

During the latter stages of the war, the Germans also forcibly conscripted men from the western regions of Poland into the Wehrmacht. This was a repetition of the situation during the First World War where Poles were forced to shoot at each other. As a boy, my father had told me about how he had been conscripted into the Tsarist Russian Army. He was sent into combat against the Austrians on the Galician Front. Knowing that there were many Poles in the Austrian ranks, he surrendered at the earliest opportunity. While there was little to recommend about spending years as a prisoner, he did manage to remain alive and in one piece until the end of the war. After all, in whose name was he to kill and spill the blood of his countrymen? Certainly not that oppressor of Poland, Tsar Nicholas II.

The situation during my first time in combat was something similar, but with one difference. While we were on our way to the front to blunt the so-called German "*Blitzkrieg*," our unit was shattered by an air attack. Instead of falling into captivity, the fragments of our regiment were forced into a "rapid withdrawal," which meant that we ran for our lives in panic. Was it fate or the finger of God that kept me from joining the over one million Red Army men taken prisoner by the Germans? Only a handful would live to see the end of the war, only then to be deported to the Gulag.

In the end, I survived. I was also was most fortunate enough to leave the "Soviet paradise" to live in freedom with my body and soul still intact. Despite intense efforts on my part, I was never able to find anyone who had served with me in the Red Army, or those who shared my experiences in the labor camps; not in Persia, nor the Middle East, or anyplace else before my arrival in Scotland. We were people scattered throughout the Polish Armed Forces, and shared a common misery, and indeed, there were many of us.

1940 - 1941
Oh, Poland Farewell!
Beloved Country,
The Bolsheviks Send You
To Their "Paradise"
Where Instead of Angels
Devils Are In Charge
And For You Poles - -
They Have Readied Graves